The Afterlife of the Leiden Anatomical Collections

The Afterlife of the Leiden Anatomical Collections starts where most stories end: after death. It tells the story of thousands of body parts kept in bottles and boxes in nineteenth-century Leiden – a story featuring a struggling medical student, more than one disappointed anatomist, a monstrous child, and a glorious past. Hieke Huistra blends historical analysis, morbid anecdotes, and humour to show how anatomical preparations moved into the hands of students and researchers, and out of the reach of lay audiences. In the process, she reveals what a centuries-old collection can teach us about the future fate of the biobanks we build today.

Hieke Huistra is an assistant professor in the history of science and medicine at Utrecht University.

The History of Medicine in Context

Series Editors: Andrew Cunningham (Department of History and Philosophy of Science, University of Cambridge) and Ole Peter Grell (Department of History, Open University)

Titles in the series include

For more information about this series, please visit: https://www.routledge.com/The-History-of-Medicine-in-Context/book-series/HMC

The Afterlife of the Leiden Anatomical Collections

Hands On, Hands Off

Hieke Huistra

LONDON AND NEW YORK

First published 2019
by Routledge
2 Park Square, Milton Park, Abingdon, Oxon OX14 4RN

and by Routledge
605 Third Avenue, New York, NY 10017

First issued in paperback 2021

Routledge is an imprint of the Taylor & Francis Group, an informa business

Publisher's Note
The publisher has gone to great lengths to ensure the quality of this reprint but points out that some imperfections in the original copies may be apparent.

British Library Cataloguing in Publication Data
A catalogue record for this book is available from the British Library

Library of Congress Cataloging in Publication Data
Names: Huistra, Hieke, 1982- author.
Title: The afterlife of the Leiden anatomical collections : hands on, hands off / by Hieke Huistra.
Description: New York : Routledge, 2019. |
Series: The history of medicine in context | Includes bibliographical references and index. |
Identifiers: LCCN 2018030660 (print) | LCCN 2018043011 (ebook) |
ISBN 9781315586144 | ISBN 9781472461070 | ISBN 9781472461070 (hbk) |
ISBN 9781315586144 (ebk)
Subjects: LCSH: Anatomical museums–Netherlands–Leiden–
History–19th century. | Human anatomy–Netherlands–Leiden–
History–19th century. |
Human anatomy–Netherlands–Leiden–Methodology–History–
19th century.
Classification: LCC QM51.N42 (ebook) | LCC QM51.N42 L38 2019
(print) | DDC 612.0074/49238–dc23
LC record available at https://lccn.loc.gov/2018030660

ISBN 13: 978-1-03-209458-8 (pbk)
ISBN 13: 978-1-4724-6107-0 (hbk)

Typeset in Sabon
by Taylor & Francis Books

Foar heit

Contents

Figures

Acknowledgements

This book started its life at Leiden University, as a PhD dissertation for the research project 'Cultures of Collecting: The Leiden Anatomical Collections in Context', funded by the Netherlands Organisation for Scientific Research (NWO) under grant 360-55-070. I thank my dissertation supervisors, Rina Knoeff and Rob Zwijnenberg, for their sharp criticism and sound advice, and for asking, time and again, what it was, exactly, that I wanted to say (Rina) and why this should be said at all (Rob). Marieke Hendriksen, my fellow PhD candidate, started her dissertation three months before I did. Without her, it would have taken me much longer to find my way around the project, the department, and the Leiden anatomical collections.

During my research, I made as many behind-the-scenes visits to anatomical institutions as I could. Without exception, these were great fun. They also proved invaluable in writing this book. I would like to thank the following people for showing me around their museums, collections, and labs: the LUMC museum committee, Gemma Angel, Tiemen Cocquyt, Martyn Cooke, Andrew Cornell, Bill Edwards, John Le Grand, Peter Heller, Fred van Immerseel, Maggie Reilly, Laurens de Rooy, and Marco de Ruiter. Special thanks to Dries van Dam and Bas Wielaard, who guided me through the archives, store rooms, and display cases of the Leiden anatomical museum, always with great enthusiasm. In addition to anatomical institutions, I visited libraries and archives; I encountered helpful staff in all of them.

Many others have helped me shape my ideas – by asking questions, by commenting on parts of the manuscript, by responding to my queries, by correcting my mistakes, or simply by nodding encouragingly. I cannot name all of them, but I would like to acknowledge at least some: Sam Alberti, Minie Baron, Klaas van Berkel, Harm Beukers, Timo Bolt, Jenny Boulboullé, Jonna Brenninkmeijer, Tatjana Buklijas, Andrew Cunningham, Mechthild Fend, Martha Fleming, Eulàlia Gassó I Miracle, Elizabeth Hallam, Nick Hopwood, Eric Jorink, Frans van Lunteren, Larissa Mendoza Straffon, Willem Otterspeer, Reina de Raat, Ruth Richardson, Ida Stamhuis, Fenneke Sysling, Ruben Verwaal, Andreas Weber, Daan Wegener, Martin Weiss, Robert-Jan Wille, Dick Willems, and Huib Zuidervaart. In 2009 and

2012, our group in Leiden organized a workshop and a conference, respectively, on anatomical collections; I am grateful to all the participants for the many fruitful discussions we had.

I carried out part of my research in London, as one of the last visiting researchers at the now-closed Wellcome Trust Centre for the History of Medicine at UCL. I thank the following people for their hospitality: Simon Chaplin, Louise King and her staff, and Agri and Roger Ford. The Leiden University Centre for the Arts in Society, Stichting Historia Medicinae, and the Leids Universiteits Fonds financed research trips and conference visits.

I finished this book at Utrecht University, surrounded by kind, intelligent, and stimulating colleagues. I am especially grateful to Bert Theunissen, who helped me find the time and space needed to complete the writing process.

One of the best things in writing this book has been sharing every step of the way with my brother, Pieter Huistra. I cherish our conversations on nineteenth-century history, twenty-first-century academia, and life in general, and I am glad that Tessa Lobbes now regularly joins these conversations.

I wish my father could have read this book in its published form. He and my mother gave me everything a child could wish for: a garden, a library, and endless support and encouragement. Thinking of them never fails to make me feel safe.

My final thanks go to my husband, Thijs Kinkhorst, and my daughter, Marte Aline. Marte arrived when this book was almost done, and managed, in her own cheerful way, to delay it a bit more. I am grateful to her for showing me how delays can be happy things as well. Thijs was there the whole time, providing food, laughter, and perspective. He made this book possible.

Introduction

In this book, death is not the end. It is not even the beginning – we enter the story weeks, months, years after the body went cold. Anatomists have already cut open the corpse, taken out the organs, injected the vessels, boiled the bones, put the resulting preparations in jars and boxes, and added these preparations to their institutions' collections. We are in the nineteenth century, the age in which institutional medical collections flourished, but we could observe similar practices in later periods, including our own. The body parts kept and the preparations made would be different, though. Nineteenth-century collections contained injected organs, macerated skeletons, bottled fetuses, stuffed animals, and microscopic slides; nowadays, laboratories and clinics keep frozen embryos, blood samples, cell lines, preserved brains, and full bodies. The new collections have not completely replaced the old ones: many of the preparations collected by nineteenth-century anatomists were never discarded. Our medical institutions keep both historical and contemporary collections, and studying the former may help us understand a thing or two about the latter. That is why I wanted to tell the story of a nineteenth-century anatomical collection.

Collecting human body parts raises moral questions about how we should handle human tissue. Most of us agree that bodily material should not be collected and stored against the wishes of the donors or their relatives, but we disagree about how explicit and specific their consent should be. Contemporary collections of bodily material, usually called biobanks, often contain samples taken primarily to diagnose or to treat the donor, such as blood drops taken from newborns during the heel prick test.[1] Can scientists use heel prick blood collected two or three decades ago to study genetic disease markers? Or should they first locate the now grown-up donors and ask them for their permission to use their blood samples no longer just for diagnosis, but now also for research? And if scientists want to reuse the blood in a new research project a few years hence, should they again seek permission? Or does it suffice to ask donors to consent just once to all possible future research use, by giving so-called broad or blanket consent?

Most researchers using biobanks prefer broad consent models, because these maximize their freedom and minimize hassle; indeed, most biobanks ask donors for broad consent, if they explicitly ask for consent at all.[2] Yet, it is unclear whether such broad consent holds up legally.[3] Furthermore, several studies have shown that at least part of the general public (i.e., potential and actual donors) would prefer more specific types of consent, for example because they want to exclude for-profit research or controversial research methods such as stem cell research or cloning.[4] Thus, as a society, we still need to determine whether we consider broad consent sufficient for future research use. And as individuals, as long as biobanks continue to use broad consent, we need to decide whether we want to consent to future research use when asked. Both questions would be easier to answer if we knew more about this future research use, but the whole point of broad consent, of course, is that we don't. We cannot predict what researchers will do with our bodily material in the future. We can, however, try to make an educated guess, and to do so, it helps to study the past. A historical case study provides the timescale we need to see what happens with collections of bodily material in the decades or centuries after they have been established. This book presents such a case study: the anatomical collections in nineteenth-century Leiden.

The Leiden collections are the oldest institutional collections in Europe; the nineteenth century was a period when medical research and teaching changed profoundly – a change sometimes labelled 'the birth of modern medicine'. In this book, I show that the old collections remained relevant in the new medicine, however fundamental the changes may have been. In doing so, I propose a new way of understanding anatomical collections: as dynamic and flexible entities, meant for hands-on use, and reused time and again. Medical historians tend to see anatomical collections as static objects, intended to be classified, arranged, and admired from a distance. By contrast, I argue that the collections were not just for looking, but also for handling: preparations were taken out of their jars, passed around to be felt and smelled in class, and reinvestigated and redissected in the laboratory. During this use, the preparations changed both physically and conceptually. For example, researchers might cut up macroscopic preparations to investigate their microscopic structure and thus adapt old preparations to new theories. This flexibility characterizes preparations (anatomical models, for instance, lack it); it follows from their material properties. As philosopher of biology Hans-Jörg Rheinberger has observed, preparations are made of what they represent.[5] This allowed researchers to return to the original tissues time and again, extracting new information that enabled reinterpretation and thus the collections' prolonged use in research and teaching. This prolonged use was further stimulated by the scarcity of raw material for the preparations: fresh human bodies were rare and their arrival was unpredictable. This encouraged researchers to retain and reuse old preparations.

Flexibility and scarcity shaped not only the path of the Leiden collections, but also the trajectories of collections in other places and periods. All preparations are made of what they represent; the raw material is rarely abundant. Indeed, as I will show with examples from other countries, old collections continued to be used in the new medicine throughout Europe.

This lifts the Leiden story above the local, and allows us to tie the past to the present. Cell lines and injected kidney preparations are made with different techniques, but both are made of what they represent, and of material that took considerable effort to acquire. Based on my analysis of the Leiden case, I will propose that these similarities make it likely that the collections we are currently building await a comparable fate to those of earlier centuries: medical institutions will hold on to them for prolonged use in research and teaching. Furthermore, I will suggest that the combination of scarcity and flexibility may also tempt researchers to hoard bodily material when they can, sometimes even without any consent. This happened, for example, in the Royal Liverpool Children's Hospital, locally known as Alder Hey; in 1999, it was discovered that doctors at the hospital had routinely been removing organs from deceased children without their parents' consent for decades.[6]

Thus, history can show us where we may be going. But it also sheds light on how we got to where we are now, and this helps us to think about the present-day dilemmas posed by historical collections. As mentioned above, many historical anatomical collections are still housed in hospitals and laboratories. Their presence raises several problems; I wish to discuss two of these.

First, according to our moral norms (and sometimes according to historical norms, too), some of the body parts in these collections should never have been collected at all. Collecting without consent was common, and some collectors even ignored the explicit instructions of the deceased or their relatives. The eighteenth-century skeleton of Charles Byrne (1761–1783), known as the Irish Giant, is a well-known example. Having been exhibited as a curiosity during his lifetime, Byrne specifically requested burial at sea to prevent his body being dissected after death, but the surgeon-anatomist John Hunter bribed his way into acquiring Byrne's body. The skeleton is still on display in the Hunterian Museum at the Royal College of Surgeons of England. Some argue that Byrne's last wish should be granted; others favour keeping the skeleton at the Hunterian, partly on the grounds that retaining it could provide us with useful medical knowledge.[7]

Another painful practice was the gathering of bones and body parts in areas colonized by the collectors' countries, as European physical anthropologists did in the nineteenth and early twentieth centuries.[8] The anthropologists used this material to develop theories that we now consider racist, as they assumed a hierarchy within the human species. This hierarchy ranked the collected below the collectors, thus offering a scientific justification for the power relations that had led to the collecting in the first place. Although the theories lost ground in the twentieth century, the material lingered in medical institutions. Since the late twentieth century, the families and groups from which the material was originally taken have increasingly requested its return, but their requests have sometimes encountered resistance. If we view the collections as the leftovers of obsolete theories, this resistance is hard to justify: if the material no longer serves any medical purpose, why not return it?

Again, the promise of future usefulness partly explains why medical institutions have been, and continue to be, reluctant to part with the material.[9] This explanation might lead us to accept their reluctance, but it might also help us to find ways to overcome it.

A second problem raised by the historical collections kept in medical institutions concerns their public accessibility. Some argue that the preparations should be put on public display. Historical anatomical preparations can teach us about two major parts of our identity: our bodies and our past. They can do so in a powerful way, because they are 'the real thing' (although simultaneously they are not, as we will see in Chapter 2). They not only display the human body, they *are* the human body. And rather than merely illustrating a historical story, they *are* from the past – they might be the closest we will ever get to meeting our ancestors. But their 'realness' also complicates displaying them. They were created out of actual people, and in many cases, we do not know whether these people consented to keeping, let alone openly displaying, their remains. Thus, the accessibility question has an ethical dimension. And yet, if we look more closely, we see that historical anatomical collections are often difficult to access not because of principled decisions, but for practical reasons: they are housed on the outskirts of towns and open on weekdays only. This book shows that these practical hurdles result from a historical process that started not because moral attitudes to displaying human body parts changed, but because medical researchers and teachers, who continued to use the collections, moved to new spaces and adopted new methods. This does not mean that ethical questions were never considered, but we should not overestimate their importance.

Acknowledging the more mundane contingencies involved in closing off anatomical collections clarifies the debate on whether they should be publicly accessible nowadays. Not only does it explain why many modern collections are closed, but it also prevents us from misinterpreting each closed collection as a moral objection to displaying human remains. Furthermore, it suggests that if we decide to open up historical anatomical collections further, we need to do more than simply allow the general public to enter: being open in theory is not the same as being accessible in practice.

The story of the Leiden anatomical collections matters, because it helps us handle both historical and contemporary anatomical collections: not by answering the ethical questions involved, but by providing a long-term perspective that is crucial for building the nuanced understanding that should, I think, precede moral judgment. The story of the Leiden anatomical collections matters for another reason as well: it helps us solve an open question in the history of medicine. In the next section, I will outline this question and explain why we should search its answer in Leiden. Subsequently, I define some of the main terms used and outline the structure of this book; after which we are ready to move on, or rather back, to nineteenth-century Leiden, where the actual story begins.

Anatomical collections in the nineteenth century

In the history of medicine, the nineteenth century is famous for two things: the birth of the clinic and the rise of the laboratory.[10] However, it was just as much the age in which institutional anatomical collections flourished; a fact that has long been overlooked by medical historians focusing on the aforementioned birth and rise. Until the early twenty-first century, most historians at best neglected nineteenth-century anatomical collections; at worst, they explicitly stated that anatomical collections had become redundant and had been replaced by hands-on learning, clinical teaching, and laboratory research.[11] In the last 15 years, historians have started to rewrite this narrative: scholars such as Erin McLeary, Samuel Alberti, and Jonathan Reinarz have shown that rather than disappearing, anatomical collections were used in medical research and teaching throughout the nineteenth century.[12] However, their insights have yet to reach general histories of modern medicine.[13] The idea that the clinic and the laboratory superseded collections and museums continues to persist, perhaps because although the flourishing of anatomical collections has been described, it has not yet been fully explained. This book extends our current explanation and in doing so aims to help promote insight that anatomical collections continued to matter in what was supposedly the age of the clinic and the laboratory.

The problem is that at first sight, anatomical collections do not seem to fit into the spaces that, according to most existing historiography, occupied centre stage in the new medicine: the clinic and the laboratory. In these new spaces, practices such as bedside teaching, dissecting, hands-on training, and experimenting played a key role. Since we are used to seeing anatomical collections as hands-off, static objects, it is easy to assume that they lost functionality: preparations, we think, were not to be touched, handled, dissected or experimented upon. But this assumption is at odds with the boom in anatomical collections during exactly this period. In universities, hospitals, and laboratories throughout Europe, large amounts of time, money, and space were invested in keeping and extending anatomical collections. In Strasbourg, for example, a small university collection of just over 200 preparations was extended rapidly from 1804 onwards: in 1820, the collection contained about 3,000 preparations; in 1870, over 4,000.[14] In Berlin, pathologist Rudolf Virchow (1821–1902) built a collection of over 23,000 preparations.[15] In London, comparative anatomist Robert Edmond Grant brought together around 10,000 preparations – and that is not counting the ones he had to discard because they had been damaged by two of the museum curator's greatest enemies, rats and students.[16]

How can we not only acknowledge, but also understand this flourishing of anatomical collections? So far, the standard explanation has been that the clinic and the laboratory were not the only spaces that mattered in the new medicine, but that museums also continued to be important, or became even more so. British historian Jonathan Reinarz has suggested we rename the nineteenth century 'the age of museum medicine'.[17] In his work on ways of knowing, historian of science and medicine John Pickstone has argued that a series of research fields emerged early in the nineteenth century that relied on a museological way of knowing, based on collecting, classifying, arranging, and comparing.[18] Some of these new fields, such as pathological anatomy and a specific type of comparative anatomy, belonged to medicine, and thus nineteenth-century medicine required museums and their collections.

Seeing the rise of the laboratory and the birth of the clinic as an addition to, instead of a replacement for, the museum partly explains the enduring importance of anatomical collections. Similar claims have been made for fields other than medicine. Historian of biology Lynn Nyhart has written about the claim that zoological laboratories replaced natural history museums at the end of the nineteenth century, and why it is wrong.[19] She suggests that the development of biology is best represented not by a tree (with natural history and its museums as the trunk, and specialties such as zoology and its laboratories as the branches), but by a growing landscape of research fields. The boundaries between the fields may change as the landscape expands, but the fields continue to coexist. Applying the landscape metaphor to the development of medicine captures how laboratories and clinic emerged as new areas, while museums did not disappear. But this is only part of the story: collections did not just exist in museums adjacent to the clinic and the lab; they were also housed *inside* the new spaces. The Leiden physiological laboratory, for instance, housed a collection, although the new physicalist orientation transformed physiology into a discipline based on an experimental way of knowing.[20]

Pickstone briefly addresses this issue, explaining that experimental styles of biology and medicine were based on data that had to be collected and stored.[21] However, it is not clear why these data collections also had to include collections of anatomical preparations, let alone collections of preparations partly created in the 'old' medicine. And yet they did, as I will show in the first two chapters with examples from Leiden and beyond. I focus on the Leiden collections because they are an excellent case not only for proving that old collections were used in new spaces, but also for investigating how this was possible, because the Leiden collections were probably the least likely collections to fit the new theories and practices. They were the oldest institutional collections in Europe, established in the late sixteenth century and in their heyday in the eighteenth century – if they could be used in the new nineteenth-century medicine, all European collections could. In other words, if we can find an explanation that covers this extreme case, it likely covers other, more moderate cases as well.

Nonetheless, it is never wise to generalize based on a single case study. One could object to taking the oldest institutional collections as the main case that perhaps the institution housing these collections was encumbered by its long past and had become old-fashioned. Maybe Leiden continued to use its collections not because they were useful in the new medicine, but because the new medicine never reached Leiden. It is true that Leiden was certainly not the first to adopt the new hands-on, experimental theories and practices, but, as we will see, the new medicine did eventually arrive, although perhaps later than elsewhere. Furthermore, to show that Leiden indeed represents a larger pattern regarding the continued use of anatomical collections in the new medicine, I will supplement the Leiden story with examples from other European cities, including institutions that were seen as front-runners in nineteenth-century medicine, such as the Royal College of Surgeons of England in London, the collections in Paris, and collections at the German universities. The general character of the Leiden case is strengthened further by the fact that, as mentioned above, my explanation – seeing anatomical collections as dynamic and flexible – relies strongly on the material characteristics of the preparations, which are shared by preparations in other places and periods. Thus, a detailed study of the Leiden collections supplemented with international examples allows me to extend the existing explanation for the enduring relevance of anatomical collections in the nineteenth century. I show that anatomical collections mattered in the new medicine, not just because the clinic and the laboratory complemented rather than surpassed the museum, but also because the collections were flexible, dynamic entities that functioned as well in the new theories and practices of the clinic and the laboratory as they did in the old practices of the museum.

Museums, collections, cabinets: on terminology

Although the terms 'museum' and 'collection' are sometimes used interchangeably, they refer to different things. In the nineteenth century, many collections existed outside of museums, and some of the institutions that were known as museums did not own collections. As explained above, to understand the enduring relevance of anatomical collections in the new medicine, we have to consider their use in spaces outside the museum. Unfortunately, this use is obscured because the terms 'museum' and 'collection' are often used interchangeably. The first step in understanding how collections functioned outside the museum is therefore to refine our use of these two words. This is tricky, not least because nineteenth-century actors often employed these words ambiguously. To complicate matters further, the Dutch terms used in Leiden do not always possess perfect parallels in English. To avoid confusion between my analytical concepts and the actors' categories, I will use this section to discuss the key words in our story: collection (*collectie* or *verzameling* in Dutch), museum (*museum*), cabinet (*kabinet*), anatomical (*anatomisch*), preparation (*preparaat*), and specimen (*specimen*). I will explain how these words were used in the nineteenth century and how I use them in my analysis.

In nineteenth-century Leiden, four different words were used to describe the assemblages of anatomical preparations and the buildings and institution that housed them: *verzameling, collectie, kabinet*, and *museum*. *Collectie* and *verzameling* were synonyms; both words translate as 'collection'. The two words were used to describe the accumulations of anatomical preparations that had been gathered by the university or by individual anatomists; *verzameling* was used most often.

I understand the analytical concept of a 'collection' as a large number of objects gathered and kept together. The objects have been selected because they possess a certain value – perhaps they are of artistic or historical importance, perhaps they serve a scientific purpose, or perhaps they are rare. The reasons for gathering objects can be many. It is hard to determine when exactly a set of objects becomes large enough to be identified as a collection, but most medical researchers or institutions gathering anatomical preparations have gone so far over the line that we can all agree that they have crossed it, even if we are not sure when exactly this happened. The more than 23,000 preparations brought together by Rudolf Virchow constituted a collection, as did the approximately 8,000 preparations housed in the mid-nineteenth-century Leiden Anatomical Cabinet.

When one has to store and work with thousands of objects, it is easy to lose track. To avoid becoming overwhelmed, collection owners and curators usually impose a structure on the things they have collected. This is especially important when collections are to be shared, as is the case with institutional collections, the main type of scientific collection from the nineteenth century onwards.[22] To facilitate the use of their collections, owners and curators have to arrange, classify, and catalogue the individual objects, and this changes the objects. Arrangements and classifications impose certain meanings on the objects they structure, and prevent these objects from assuming other meanings.[23] Sometimes, transforming the individual objects is the main aim of bringing the objects together; studying individual objects as part of a larger whole by arranging, classifying, and comparing them can produce new knowledge. In other cases, however, the change of meaning that the objects in a collection undergo is unintended, an inescapable by-product of the practical need to structure the collection somehow. This often happens with anatomical preparations. Body parts are not necessarily collected to become part of an orderly museum collection. There are more prosaic reasons for collecting them: bodily material is scarce and decays rapidly, meaning that you need to catch it while you can and then store it safely for future use. This future use may occur outside the context of the collection, such as when preparations are taken into the dissection hall or the lecture room. But even then, the stored preparations will form a collection, imposing a certain structure that changes the meaning, intentionally or unintentionally, of the preparations, and favours some interpretations above others.

Kabinet and *museum* are more ambiguous words than *collectie* and *verzameling*, and they were used inconsistently in the nineteenth century. In 1864, the *Nieuw woordenboek der Nederlandsche taal* (New dictionary of the Dutch language) defined *museum* as follows:

Museum, n. [neuter] ([pl.] ...ea), building –, institution dedicated to art or science; art cabinet, cabinet (mainly) of objects of natural history, etc.[24]

(Museum, o. (...ea), gebouw –, instelling aan kunst of wetenschap gewijd; kunstkabinet, kabinet (voornamelijk) van voorwerpen der natuurlijke historie enz.)

Museum could thus refer to a collection or the institution housing the collection (both meanings are implied in *kabinet* in the second part of the definition), but it could also mean 'building or institution dedicated to art or science'.[25] To be called a *museum*, this building or institution did not need to own a collection, nor did it need to be open to visitors. Towards the end of the nineteenth century, this meaning disappeared: in 1908, the *Woordenboek der Nederlandsche taal* (Dictionary of the Dutch language) called it 'now obsolete in our language'.[26] In the period covered by this book, however, *museum* was still regularly used in this way in the Netherlands. In England, this use seems to have disappeared even earlier, before the nineteenth century; the *Shorter Oxford English Dictionary* claims that 'museum' was last used in this way in the late eighteenth century.[27] Nonetheless, 'museum' remained an ambiguous term in nineteenth-century English: like the Dutch equivalent, it could designate both a collection and the institution housing this collection. Nowadays, 'museum' still carries this double meaning, as the definition in the *Oxford English Dictionary* (OED) reveals:

museum, n. [...] A building or institution in which objects of historical, scientific, artistic, or cultural interest are preserved and exhibited. Also: the collection of objects held by such an institution.[28]

Due to this double meaning, the term 'museum' quickly becomes confusing when discussing the relationship between museums and collections. Therefore, when I use 'museum' as an analytical category, I *never* use it to refer to a collection. Moreover, whenever possible, I use 'museum' to refer to the institution and 'museum building' to refer to the structure housing this institution.

With these modifications, I have reduced the *OED* definition to: 'an institution in which objects of historical, scientific, artistic, or cultural interest are preserved and exhibited'. Exhibition is crucial in museums; at least some of the objects in a museum are *on display*. They are meant to be observed by an audience. This audience is not necessarily 'the general public'; it may consist of students or researchers, for example, instead of lay visitors. Many museum professionals and museum studies scholars consider being open to a broad audience a key characteristic of a museum. This can be seen, for example, in the definition of a museum used by the International Council of Museums (ICOM). The latest definition, adopted in 2007, states that 'a museum is ... open to the public'.[29] Indeed, most of the present-day institutions called 'museums' are open to non-specialist visitors, but some of them, including many anatomical museums, are not, or only to a very limited extent. The Leiden Anatomical Museum offers a case in point: it is open to the general public two days a year. Other examples include the Wellcome Museum of Anatomy and Pathology at the Royal College of Surgeons of England and the Gordon Museum of Pathology at the King's College medical campus, both in London. All three institutions are called 'museums', but their main audience consists of specialist visitors (medical students and researchers); laypersons are not allowed, although exceptions are sometimes made on special occasions. In the nineteenth century, museums with restricted access were more common. The idea of a museum as an institution open to all only emerged in the nineteenth century, hence the changing meaning of the term during this century.[30] Thus, when I use the term 'museum', I refer to an institution where exhibiting takes centre stage, but the audience does not necessarily include laypersons.

Kabinet, which I translate as 'cabinet', was used even more ambiguously than *museum* in the nineteenth century. It could refer to an institution housing collections, to a building, room or cupboard in which collections were kept, or to a collection itself. All of these uses were common in Leiden. Multiple uses frequently occurred in the same text, even if this text was a national law.[31] I do not use 'cabinet' as an analytical category. The word appears in this book only in quotations from primary sources and as part of the proper name Anatomical Cabinet (*Anatomisch Kabinet*). The Anatomical Cabinet was the institution that housed the university's principal anatomical collections. It was known by many names in the nineteenth century, including Anatomical Cabinet, Anatomical Museum, Cabinet of Anatomy, and Anatomical-Physiological Cabinet. To keep things as clear as possible, I mainly use Anatomical Cabinet, sometimes shortened to Cabinet.

This brings us to another word that requires clarification: 'anatomical'. I use this term broadly, meaning that 'anatomical collections' contain not only preparations of 'general' or 'healthy' anatomy, but also of pathological and comparative anatomy, both macroscopic and microscopic. 'Comparative anatomy' can mean many things, but we will come to this later, in Chapter 2. For now, the term should be interpreted as involving the comparison of human and animal structures, which means that 'anatomical collections' contained animal preparations as well. In this book, I am primarily concerned with anatomical collections of *preparations*, not of *models*.[32] As we will see in Chapter 2, these two are by no means identical. Neither are *preparations* and *specimens*; at least, not in the nineteenth-century sense of the words. Today, 'specimen' is often used to denote prepared and preserved body parts, but in the nineteenth century this hardly ever happened. The exact use is hard to define, but roughly, 'preparations' were preserved body parts which had been created through some sort of dissection, whereas 'specimens' were objects such as stuffed animals, displaying the outside of the body.[33]

The structure of this book

This book examines how collections of human body parts function in the long run: it focuses not on the making of anatomical preparations, but on what happens to them in the decades and centuries after they have been created and collected.[34] Who uses them, in what way, how do the collections shape this use, and how does this use change the collections? These questions require multiple answers: anatomical collections, both in Leiden and elsewhere, had several audiences, and each audience used the collections differently. For this reason, each of the four chapters in this book focuses on a different audience: first students, then researchers, followed by lay visitors, and in conclusion, institutional administrators. These four were the main audiences in Leiden; other collections had similar audiences, although their relative importance differed between collections (and also, as we will see, between periods). The type of administrators involved depended on the institutional setting; in Leiden, the collections ultimately fell under the responsibility of the university governors, the so-called *curatoren*.

All audiences approached the collections with their own aims and interests. They did so actively; we should not see them as mere observers or passive recipients. They interpreted the collections, bringing their own experiences, knowledge, and world views, thereby adding new meanings to the preparations.[35] This does not mean that they could alter the collections as they saw fit: the preparations facilitated some interpretations and uses, while resisting others. In particular, as we will see, non-medical audiences (lay visitors and institutional administrators) increasingly encountered practical and conceptual hurdles when attempting to use the collections. This resulted in their ceasing to use the collections altogether in the second half of the nineteenth century, because they could no longer relate to them or present them as they wished. The collections disappeared from public view as a direct consequence of the continued use of the collections by the other two audiences: students and researchers.

The first two chapters discuss these two medical audiences. Together, these chapters flesh out the revised view of anatomical collections as dynamic and flexible entities. Chapter 1 shows how students actively handled the preparations instead of just looking at them, and how this made preparations relevant in all teaching spaces, not just in museums. Chapter 2 addresses the flexibility of the preparations. It explains how researchers not only handled preparations, but handled *the same* preparations for decades on end, by continuously reinterpreting them. The two chapters explain how students and researchers continued to use the anatomical collections throughout the nineteenth century. Thus, although the chapters discuss some of the ways in which this use changed during the century, the focus is on the continuity of use, not change. For this reason, the chapters are not based on a strict periodization. By contrast, in Chapters 3 and 4 on non-medical audiences, the nineteenth century is divided into two: before and after the collections moved into the laboratory.

Laboratories quickly gained in importance from the middle of the century onwards, first in teaching, then in research. In Leiden, the crucial year was 1860: in this year, the university's main anatomical collections were moved from the old church they had shared with the university library to a newly built educational complex housing the teaching laboratories for the natural sciences.

Chapter 3 shows that when they were moved to the new laboratories, anatomical collections ended up in locations that were difficult to approach and in arrangements that were hard to interpret without a medical background. Lay visitors found it difficult to relate to them; in Leiden, laypersons stopped visiting the Anatomical Cabinet in the second half of the nineteenth century. In Chapter 4, we see that the administrators of Leiden University also stopped using the collections. Before the 1860 move, they had employed the collections as a status symbol embodying the university's glorious past. In their new location, however, the preparations lost the link to their eighteenth-century creators and therefore their utility as a status symbol – a problem familiar to administrators elsewhere, as well as to later medical historians.

To conclude the book, I return to the question we started with: how should we deal with collections of human tissue, both historical and contemporary? I argue that, in addition to nineteenth-century anatomical collections, collections of bodily material from other times should also be understood as dynamic and flexible entities. Drawing on my analysis of what happened in nineteenth-century Leiden, I investigate how this understanding could inform present-day debates on handling collections of human body parts. First, though, let us leave the present time and travel back to the nineteenth century to meet our first preparation: a severed head.

Notes

1 On storage of blood collected with the heel prick test, see for example Tybjerg, 'Bottled Babies', 271 (on Denmark); Geesink and Steegers, *Nader gebruik*, 30–31 (on the Netherlands).
2 Master, Campo-Engelstein, and Caulfield, 'Scientists' Perspectives'; Edwards et al., 'Biobanks Containing Clinical Specimens'; Master et al., 'Biobanks'.
3 Caulfield and Murdoch, 'Genes, Cells, and Biobanks'.
4 Caulfield and Murdoch.
5 Rheinberger, 'Präparate'; Rheinberger, *Epistemology of the Concrete*, 232–43.
6 Redfern, Keeling, and Powell, *The Royal Liverpool Children's Inquiry*.
7 See for example Smith et al., 'We Cannot Change'.
8 In addition to bodily material, anthropologists also collected bodily measurements, another past practice raising ethical objections nowadays. See for example Sysling, *Racial Science*.
9 Fabian, *The Skull Collectors*, 218. Public museums also use possible scientific value as an argument to refuse transfer of (colonial) human remains. See for example The British Museum Policy on Human Remains in the Collection, article 5.2.2 (p. 3), available on www.britishmuseum.org/pdf/Human%20Remains%20policy%20July%202013%20FINAL.pdf, accessed 20 December 2017.
10 Ackerknecht, *Paris Hospital*; Foucault, *Birth of the Clinic*; Cunningham and Williams, *Laboratory Revolution*.

11 For an example of the latter, see Wachelder, *Universiteit*, 100–102, on the Dutch case.

12 McLeary, 'Science in a Bottle'; Alberti, *Morbid Curiosities*; Reinarz, 'Age of Museum Medicine'. On nineteenth-century anatomical collections, see also Alberti and Hallam, *Medical Museums*; Burmeister, 'Popular Anatomical Museums'; Close Koenig, 'Betwixt and Between'; Fröber, 'Anatomical Collection in Jena'; Hallam, *The Anatomy Museum*; Knoeff and Zwijnenberg, *Fate*; Matyssek, *Rudolf Virchow*; Redman, *Bone Rooms*; Sappol, 'Morbid Curiosity'; Turnbull, *Collecting the Indigenous Dead*. On anatomical collections before and after the nineteenth century, see for example Chaplin, 'John Hunter'; Hendriksen, *Elegant Anatomy*; Jones, 'The Mütter Museum'; Margócsy, 'Commercial Exchange'; Morgan, *Icons of Life*; Schultka and Neumann, *Anatomie*.

13 See for example Bynum et al., *Western Medical Tradition*; Bynum and Porter, *Companion Encyclopedia*; Jackson, *Oxford Handbook*.

14 Close Koenig, 'Betwixt and Between', 76–77.

15 Matyssek, *Rudolf Virchow*, 31.

16 Parker, *Robert Edmond Grant*, 17.

17 Reinarz, 'Age of Museum Medicine'.

18 Pickstone, 'Museological Science?'

19 Nyhart, 'Natural History', 435–42.

20 Cunningham, 'Old Physiology'; Nyhart, *Biology Takes Form*, 67–80.

21 Pickstone, *Ways of Knowing*, 75–76.

22 Alberti, 'Owning and Collecting'.

23 For a theoretical analysis of what it means for an object to belong to a collection, see for example Pearce, *Museums, Objects and Collections*.

24 Calisch and Calisch, *Nieuw woordenboek*, 813.

25 For lack of a better word, I use 'science' to translate the Dutch *wetenschap*, although the latter has a broader meaning, similar to the German *Wissenschaft*.

26 *Woordenboek der Nederlandsche taal*, s.v. 'museum', accessed 28 December 2017, http://gtb.inl.nl.

27 *Shorter Oxford English Dictionary*, 6th ed., s.v. 'museum'. On early modern use of the term 'museum' and its French equivalent 'muséum', see Findlen, 'The Museum'; Lee, 'Musaeum of Alexandria'.

28 *OED Online*, s.v. 'museum', accessed 28 December 2017, www.oed.com/view/Entry/124079.

29 ICOM website, 'Museum definition', definition adopted in 2007, accessed 13 January 2017, http://icom.museum/the-vision/museum-definition/.

30 Bennett, *Birth of the Museum*.

31 See for example 'Organiek Besluit Hooger Onderwijs', 2 August 1815.

32 On anatomical models, see Alberti, 'Wax Bodies'; Dacome, *Malleable Anatomies*; Ebenstein, *The Anatomical Venus*; Hopwood, *Embryos in Wax*; Hopwood, 'Artist versus Anatomist'; Maerker, *Model Experts*; Mazzolini, 'Plastic Anatomies'; Messbarger, *The Lady Anatomist*; Pirson, *Corps à corps*; Schnalke, 'Casting Skin'.

33 See Chaplin, 'John Hunter', 101–2 on the early modern use of both terms, which closely resembles nineteenth-century practices.

34 On the lives of objects after they enter a collection, see also Alberti, 'Objects and the Museum'.

35 In recent history of science, audiences are usually understood as such active users – not just collection audiences but also audiences of, for example, books, scientific instruments, or theories. See also Secord, 'Knowledge in Transit'. Specifically on understanding (lay) audiences of anatomical collections as active users, see Alberti, 'The Museum Affect'; Knoeff, 'The Visitor's View'.

Bibliography

Ackerknecht, Erwin H. *Medicine at the Paris Hospital, 1794–1848*. Baltimore: Johns Hopkins University Press, 1967.

Alberti, Samuel J. M. M. 'Objects and the Museum'. *Isis* 96(2005): 559–71. https://doi.org/10.1086/498593.

Alberti, Samuel J. M. M. 'Owning and Collecting Natural Objects in Nineteenth-Century Britain'. In *From Private to Public: Natural Collections and Museums*, edited by Marco Beretta, 141–54. Sagamore Beach: Science History Publications, 2005.

Alberti, Samuel J. M. M. 'The Museum Affect: Visiting Collections of Anatomy and Natural History'. In *Science in the Marketplace: Nineteenth-Century Sites and Experiences*, edited by Aileen Fyfe and Bernard Lightman, 371–403. Chicago: University of Chicago Press, 2007.

Alberti, Samuel J. M. M. 'Wax Bodies: Art and Anatomy in Victorian Medical Museums'. *Museum History Journal* 2(2009): 7–36.

Alberti, Samuel J. M. M. *Morbid Curiosities: Medical Museums in Nineteenth-Century Britain*. Oxford: Oxford University Press, 2011.

Alberti, Samuel J. M. M., and Elizabeth Hallam, eds. *Medical Museums: Past, Present, Future*. London: Royal College of Surgeons of England, 2013.

Bennett, Tony. *The Birth of the Museum: History, Theory, Politics*. London: Routledge, 1995.

Burmeister, Maritha Rene. 'Popular Anatomical Museums in Nineteenth-Century England'. PhD diss., Rutgers University, 2000.

Bynum, William F., Anne Hardy, L. Stephen Jacyna, Christopher Lawrence, and E. M. Tansey. *The Western Medical Tradition, 1800 to 2000*. Cambridge: Cambridge University Press, 2006.

Bynum, William F., and Roy Porter, eds. *Companion Encyclopedia of the History of Medicine*. 2 vols. London: Routledge, 1993.

Calisch, Isaac Marcus, and Nathan Salomon Calisch. *Nieuw woordenboek der Nederlandsche taal*. Tiel: Campagne, 1864. www.dbnl.org/tekst/cali003nieu01_01/index.php.

Caulfield, Timothy, and Blake Murdoch. 'Genes, Cells, and Biobanks: Yes, There's Still a Consent Problem'. *PLOS Biology* 15, no. 7(2017): e2002654. https://doi.org/10.1371/journal.pbio.2002654.

Chaplin, Simon. 'John Hunter and the "Museum Oeconomy", 1750–1800'. PhD diss., University of London, 2009.

Close Koenig, Tricia. 'Betwixt and Between: Production and Commodification of Knowledge in a Medical School Pathological Anatomy Laboratory in Strasbourg (Mid-19th Century to 1939)'. PhD diss., Université de Strasbourg, 2011. http://scd-theses.u-strasbg.fr/2323.

Cunningham, Andrew. 'The Pen and the Sword: Recovering the Disciplinary Identity of Physiology and Anatomy before 1800. I: Old Physiology – the Pen'. *Studies in History and Philosophy of Biological and Biomedical Sciences* 33(2002): 631–65. https://doi.org/10.1016/S1369-8486(02)00023-7.

Cunningham, Andrew, and Perry Williams, eds. *The Laboratory Revolution in Medicine*. Cambridge: Cambridge University Press, 1992.

Dacome, Lucia. *Malleable Anatomies: Models, Makers, and Material Culture in Eighteenth-Century Italy*. Oxford: Oxford University Press, 2017.

Ebenstein, Joanna. *The Anatomical Venus*. London: Thames and Hudson, 2016.

Edwards, Teresa, R. Jean Cadigan, James P. Evans, and Gail E. Henderson. 'Biobanks Containing Clinical Specimens: Defining Characteristics, Policies, and Practices'. *Clinical Biochemistry* 47(2014): 245–51. https://doi.org/10.1016/j.clinbiochem.2013.11.023.

Fabian, Ann. *The Skull Collectors: Race, Science, and America's Unburied Dead*. Chicago: University of Chicago Press, 2010.

Findlen, Paula. 'The Museum: Its Classical Etymology and Renaissance Genealogy'. *Journal of the History of Collections* 1(1989): 59–78.

Foucault, Michel. *The Birth of the Clinic: An Archaeology of Medical Perception*. Translated by A. M. Sheridan. London: Tavistock, 1976.

Fröber, Rosemarie. 'The Anatomical Collection in Jena and the Influence of Carl Gegenbaur'. *Theory in Biosciences* 122(2003): 148–61.

Geesink, Ingrid, and Chantal Steegers. *Nader gebruik nader onderzocht: Zeggenschap over lichaamsmateriaal*. The Hague: Rathenau Instituut, 2009. www.rathenau.nl/nl/publicatie/nader-gebruik-nader-onderzocht-zeggenschap-over-lichaamsmateriaal-0.

Hallam, Elizabeth. *The Anatomy Museum: Death and the Body Displayed*. London: Reaktion Books, 2016.

Hendriksen, Marieke M. A. *Elegant Anatomy: The Eighteenth-Century Leiden Anatomical Collections*. History of Science and Medicine Library 47. Leiden: Brill, 2015.

Hopwood, Nick. *Embryos in Wax: Models from the Ziegler Studio, with a Reprint of 'Embryological Wax Models' by Friedrich Ziegler*. Cambridge: Whipple Museum of the History of Science, 2002.

Hopwood, Nick. 'Artist versus Anatomist, Models against Dissection: Paul Zeiller of Munich and the Revolution of 1848'. *Medical History* 51(2007): 279–308. https://doi.org/10.1017/S0025727300000144.

Jackson, Mark, ed. *The Oxford Handbook of the History of Medicine*. Oxford: Oxford University Press, 2013. First published 2011.

Jones, Nora L. 'The Mütter Museum: The Body as Spectacle, Specimen, and Art'. PhD diss., Temple University, 2002.

Knoeff, Rina. 'The Visitor's View: Early Modern Tourism and the Polyvalence of Anatomical Exhibits'. In *Centres and Cycles of Accumulation In and Around the Netherlands*, edited by Lissa Roberts, 155–76. Berlin: Lit Verlag, 2011.

Knoeff, Rina, and Robert Zwijnenberg, eds. *The Fate of Anatomical Collections*. The History of Medicine in Context. Farnham: Ashgate, 2015.

Lee, Paula Young. 'The Musaeum of Alexandria and the Formation of the *Muséum* in Eighteenth-Century France'. *Art Bulletin* 79(1997): 385–412. www.jstor.org/stable/3046259.

Maerker, Anna. *Model Experts: Wax Anatomies and Enlightenment in Florence and Vienna, 1775–1815*. Manchester: Manchester University Press, 2011.

Margócsy, Dániel. 'A Museum of Wonders or a Cemetery of Corpses? The Commercial Exchange of Anatomical Collections in Early Modern Netherlands'. In *Silent Messengers: The Circulation of Material Objects of Knowledge in the Early Modern Low Countries*, edited by Sven Dupré and Christoph Lüthy, 185–215. Berlin: Lit Verlag, 2011.

Master, Zubin, Lisa Campo-Engelstein, and Timothy Caulfield. 'Scientists' Perspectives on Consent in the Context of Biobanking Research'. *European Journal of Human Genetics* 23(2015): 569–74. https://doi.org/10.1038/ejhg.2014.143.

Master, Zubin, Erin Nelson, Blake Murdoch, and Timothy Caulfield. 'Biobanks, Consent and Claims of Consensus'. *Nature Methods* 9(2012): 885–88. https://doi.org/10.1038/nmeth.2142.

Matyssek, Angela. *Rudolf Virchow, das pathologische Museum: Geschichte einer wissenschaftlichen Sammlung um 1900*. Schriften aus dem Berliner Medizinhistorischen Museum 1. Darmstadt: Steinkopff, 2002.

Mazzolini, Renato G. 'Plastic Anatomies and Artificial Dissections'. In *Models: The Third Dimension of Science*, edited by Soraya Chadarevian and Nick Hopwood, 43–70. Stanford: Stanford University Press, 2004.

McLeary, Erin Hunter. 'Science in a Bottle: The Medical Museum in North America, 1860–1940'. PhD diss., University of Pennsylvania, 2001.

Messbarger, Rebecca. *The Lady Anatomist: The Life and Work of Anna Morandi Manzolini*. Chicago: University of Chicago Press, 2010.

Morgan, Lynn M. *Icons of Life: A Cultural History of Human Embryos*. Berkeley: University of California Press, 2009.

Nyhart, Lynn K. *Biology Takes Form: Animal Morphology and the German Universities, 1800–1900*. Chicago: University of Chicago Press, 1995.

Nyhart, Lynn K. 'Natural History and the "New" Biology'. In *Cultures of Natural History*, edited by Nicholas Jardine, James A. Secord, and Emma C. Spary, 426–43. Cambridge: Cambridge University Press, 1996.

Parker, Sarah E. *Robert Edmond Grant (1793–1874) and His Museum of Zoology and Comparative Anatomy*. London: Grant Museum of Zoology, 2006.

Pearce, Susan M. *Museums, Objects and Collections: A Cultural Study*. Leicester: Leicester University Press, 1992.

Pickstone, John V. 'Museological Science? The Place of the Analytical Comparative in Nineteenth-Century Science, Technology and Medicine'. *History of Science* 32 (1994): 111–38.

Pickstone, John V. *Ways of Knowing: A New History of Science, Technology and Medicine*. Chicago: University of Chicago Press, 2000.

Pirson, Chloé. *Corps à corps: Les modèles anatomiques entre art et médecine*. Paris: mare & martin, 2009.

Redfern, Michael, Jean W. Keeling, and Elizabeth Powell. *The Royal Liverpool Children's Inquiry: Report*. London: The Stationery Office, 2001.

Redman, Samuel J. *Bone Rooms: From Scientific Racism to Human Prehistory in Museums*. Cambridge: Harvard University Press, 2016.

Reinarz, Jonathan. 'The Age of Museum Medicine: The Rise and Fall of the Medical Museum at Birmingham's School of Medicine'. *Social History of Medicine* 18 (2005): 419–37. https://doi.org/10.1093/shm/hki050.

Rheinberger, Hans-Jörg. 'Präparate – "Bilder" ihrer selbst: Ein bildtheoretische Glosse'. In *Oberflächen der Theorie*, 9–19. Bildwelten des Wissens: Kunsthistorisches Jahrbuch für Bildkritik, vol 1:2. Berlin: Akademie Verlag, 2003.

Rheinberger, Hans-Jörg. *An Epistemology of the Concrete: Twentieth-Century Histories of Life*. Translated by G. M. Goshgarian. Durham: Duke University Press, 2010.

Sappol, Michael. '"Morbid Curiosity": The Decline and Fall of the Popular Anatomical Museum'. *Common-Place* 4, no. 2(2004). www.common-place.org/vol-04/no-02/sappol/.

Schnalke, Thomas. 'Casting Skin: Meanings for Doctors, Artists, and Patients'. In *Models: The Third Dimension of Science*, edited by Soraya Chadarevian and Nick Hopwood, 207–41. Stanford: Stanford University Press, 2004.

Schultka, Rüdiger, and Josef N. Neumann, eds. *Anatomie und Anatomische Samm-lungen im 18. Jahrhundert*. Berlin: Lit Verlag, 2007.

Secord, James A. 'Knowledge in Transit'. *Isis* 95(2004): 654–72. https://doi.org/10.1086/430657.

Smith, Martin, Christopher Knüsel, Andrew Chamberlain, and Piers D. Mitchell. 'We Cannot Change the Past, But We Can Learn From It'. *British Medical Journal* 344(2012): e556. https://doi.org/10.1136/bmj.e556.

Sysling, Fenneke. *Racial Science and Human Diversity in Colonial Indonesia*. Singapore: NUS Press, 2016.

Turnbull, Paul. *Science, Museums and Collecting the Indigenous Dead in Colonial Australia*. Palgrave Studies in Pacific History. Palgrave Macmillan, 2017.

Tybjerg, Karin. 'From Bottled Babies to Biobanks: Medical Collections in the Twenty-First Century'. In *The Fate of Anatomical Collections*, edited by Rina Knoeff and Robert Zwijnenberg, 263–78. The History of Medicine in Context. Farnham: Ashgate, 2015.

Wachelder, Joseph. *Universiteit tussen vorming en opleiding: De modernisering van de Nederlandse universiteiten in de negentiende eeuw*. Hilversum: Verloren, 1992.

1 Remove lid before use

How students handled anatomical preparations

One of the most famous gothic stories in Dutch literature features a Leiden medical student and his finest anatomical preparation. In the depths of a stormy night, the student, as pale as the moonlight, breaks into the anatomy building. He uses a smelling bottle to revive the body of a hanged man and then steals the man's head while it – he? – is still alive. After connecting the head to an array of bottles, pouches, and wires, the student takes this living preparation to his room, where he stores it behind his bookcase. Late in the evening, when no one will come, the student takes out the head. He interrogates it, and microscopically examines the tears it sheds out of despair. One night, the head bites the student's finger, just as the police are banging on the door. The student pulls the head from the apparatus to release his finger and jumps out of the window. The head is found dead on the floor; the student is never seen again. All that is left is a mysterious book, kept in some old libraries: *Caput sedes animi: disquisitio, qua probatur artem fungi posse vice corporis, dummodo caput supersit* (The head as the seat of the soul: an investigation, with which it is proven that art can execute the duty of the body, as long as the head is still alive).[1] The author remains unknown to this day.

The story was written by the Dutch artist Alexander Verhuell (1822–1897), known for his satirical drawings of student life.[2] It first appeared in the Leiden student almanac of 1847, in blood-red ink; it has remained in print ever since.[3] Although Verhuell implied the story was set long before his time, it reflects nineteenth-century medical theories and practices and the anxieties they aroused in society, as did much gothic horror writing.[4] Verhuell and other writers, including Mary Shelley, Jules Janin, and Georges Balzac, captured the fears of their contemporaries.[5] People feared anatomists because the latter acquired bodies in ways that, especially in Britain, were murky at best, murderous at worst, and because their experiments dehumanized these bodies. Verhuell's experiment is fictitious only in that the student succeeds in keeping the head alive; several actual nineteenth-century researchers attempted to do the same, but failed.[6] Galvanists tried to resurrect bodies using electricity; mesmerists tried to communicate with the dead, something that Verhuell himself had attempted to do as well.[7] Theories like galvanism and mesmerism are reflected in Verhuell's story, as are practices such as body snatching, dissecting, and – the topic of this chapter – student handling of anatomical preparations.

Verhuell's student engaged with his treasured head just like nineteenth-century medical students worked with their preparations: in a hands-on, emotionally detached, and question-driven way, outside the medical museum. Used thus, preparations helped students learn basic anatomy, become familiar with rare pathological conditions, study phenomena invisible in fresh corpses, answer research questions for their dissertations, and get used to working with dead bodies. This chapter shows how anatomical preparations helped medical students to become better doctors. It starts with an overview of nineteenth-century medical education, which will show that medical students visited many spaces: lecture rooms, dissection halls, museums, hospitals, and laboratories. Guided by a Leiden student – proper introductions will follow shortly – we enter these spaces one by one to see how preparations assisted students' learning in all of them. The experiences of our Leiden student will be supplemented with examples of students and teachers from elsewhere in Europe and the United States. Their educational systems differed, but, as we will see, their handling of preparations was remarkably similar: lids off and hands on.

Medical education in the nineteenth century

On Tuesday 20 August 1833, the Leiden medical student Jan Bastiaan Molewater (1813–1864) started a diary.[8] He chose that particular day because he was 'in a fairly calm, diligent mood and reasonably pleased with myself' – something that did not happen very often, at least not on the days he wrote his diary.[9] Many of the entries are self-reproachful. Despite repeated resolutions, Molewater failed to get out of bed early, study as planned, and stop sleeping with the mysterious L. J. In other words, he seems to have been a typical student, and his diary may show us how nineteenth-century medical students studied medicine on the days they succeeded in getting out of bed early enough to make it to class.

Molewater's classes involved more than simply sitting and listening to professors lecturing. Nineteenth-century medical students had to use their hands, eyes, and noses in addition to their ears. This distinguished them from their predecessors at early modern universities: until the second half of the eighteenth century, formal medical teaching was mainly theoretical. Even in Leiden, little practical teaching took place. The Leiden medical faculty would later become famous for its teaching hospital, where Herman Boerhaave supposedly taught students during his rounds in the early eighteenth century. However, the hospital housed hardly any patients in Boerhaave's day, making it unlikely that he carried out much clinical teaching.[10] University medical students would treat their first patients when they started practising as physicians – a practice that in itself was hands-off, consisting mainly of listening to symptoms, deducing diagnoses, and prescribing treatment. Physicians left hands-on treatment to other healers, such as surgeons and midwives. These healers had their own, informal training methods, often apprenticeships where pupils were taught hands-on, on the job.

The shift from theory to practice in formal medical teaching, the type of teaching in which the Leiden collections functioned, occurred throughout Europe.[11] Its origins are often traced back to late eighteenth-century France, where post-revolutionary educational structures brought together surgeons and physicians, and increased the importance of hands-on practices in teaching. However, as medical historian Thomas Bonner has shown in his comparative history of modern medical education, things had already started to change before the French Revolution, both in and beyond France.[12] From 1750 onwards, surgery and medicine grew closer together in many European countries. Simultaneously, medical teachers and students explored more practical training methods. Students increasingly left lecture rooms to visit hospitals and dissection halls. They engaged with patients in the flesh instead of on the page, practiced procedures hands-on, and carried out experiments. Around 1800, the movement towards more practical teaching was well under way and it continued to grow stronger as the nineteenth century progressed. In the middle of the nineteenth century, another hands-on teaching space would be introduced: the teaching laboratory.

The ideal of practical, hands-on teaching was widely embraced, but its institutional embedding varied between countries. In Germany and the Netherlands, the universities remained central in training future physicians; in many other countries, the role of the universities decreased. In France, medical schools were often tied to hospitals, while British medical education relied heavily on private medical schools. In all countries, the state began to play a greater role in medical education and the licensing of medical practitioners, but more so on the Continent than in Britain. For example, British medical practitioners were licensed by two medical professional organizations, the Royal Colleges of Surgeons and of Physicians, until 1858, whereas in many continental countries, the state already controlled access to the medical profession by then. This difference mattered because the criteria established for medical licensing defined the objectives of medical training.

Although the organizational structures of nineteenth-century national educational systems diverged, we should not see them as strictly separated. Many students travelled abroad and followed part of their training in different educational systems. Paris was a popular destination, as were Vienna and Berlin, but students also visited smaller, less famous cities and institutions.[13] Medical historian Stephan Curtis has argued that such travels (which were undertaken not just by students but also by practitioners) contributed to the emergence of a 'European medicine': a shared body of knowledge and practices that transcended national borders.[14] Hands-on teaching in hospitals, museums, dissection rooms, and laboratories was part of this European medicine.

The travels also illustrate how students shaped their own education, which they could likewise do if they stayed at home. They could choose their own teachers, topics, and schedules, although some systems offered more freedom to do so than others. The Dutch system was highly prescriptive. Physicians had to be *medicinae doctor*, which meant that they had to go to university – unlike surgeons, pharmacists, and midwives, who were trained outside the university as well. Until 1865, a *medicinae doctor* could start practising immediately; after 1865, new legislation required an additional practical exam to be taken outside the university. University curricula were set out in the Royal Decree on Higher Education (1815), the law that regulated Dutch higher education until 1876, when the Higher Education Act was issued.[15] The Royal Decree prescribed which courses universities should offer, ensuring that Molewater's curriculum resembled that of students in Groningen and Utrecht, the other Dutch universities at the time. Students at all three universities had to satisfy the same requirements to be awarded a medical degree. They had to pass exams in anatomy, physiology, pathology, pharmacy, *materia medica*, practical medicine, surgery, and obstetrics; attend classes in natural history, comparative anatomy, dietetics, and forensic medicine; perform a dissection; participate in clinical teaching; and demonstrate their ability to interpret Hippocrates. The exact content of the classes depended on the professors teaching them, so some variation existed between universities.

The Royal Decree also stated which 'material assistance' had to be present at the universities.[16] For medical teaching, it prescribed a collection of medical books in the library; an academic hospital for clinical teaching; a collection of surgical and obstetrical instruments (both contemporary and historical); and collections of anatomical, pathological, physiological, and comparative-anatomical preparations.[17] Hospital, library, and collections were considered crucial resources; without them, according to the Decree, medical education would be 'inadequate'.[18] Medical professors agreed and urged university governors to implement these requirements. Since all universities already owned substantial libraries, most of their time and money went to building the required collections and establishing teaching hospitals (later followed by teaching laboratories, which would become mandatory in the 1876 Higher Education Act). In response to the Royal Decree, Leiden University acquired two anatomical collections, together containing almost 5,000 preparations. (Although the collections were newly acquired, the preparations they contained were not newly made; many had been created in the eighteenth century. This will be discussed in more detail in Chapter 2.) Universities elsewhere, both in the Netherlands and beyond, likewise extended their collections: just as the importance of hands-on teaching was widely agreed upon, so was the importance of anatomical collections as a material resource in medical teaching.[19] These collections were not only used in museum teaching, but they also played a role in the lecture room, the dissection hall, the laboratory, and the clinic. We will follow our student

Molewater, and some of his counterparts in other cities, through these spaces to see how preparations were integrated into the hands-on training that increasingly filled the universities.

Anatomical collections around town

Hospitals, laboratories, museums, dissection halls, and lecture rooms all take up space; hence, nineteenth-century medical teaching was spread across town, as were the collections used in it. The largest anatomical collections could usually be found in museum spaces such as the Leiden Anatomical Cabinet and the London Royal College of Surgeons' museum. Hospitals and laboratories housed collections as well. In Leiden, the academic hospital held a pathological-anatomical collection; the physiological laboratory (founded in 1866) collected microscopic slides; and the pathological-anatomical laboratory (1885) received pathological-anatomical preparations from the Cabinet.[20] The zootomical laboratory (1876) was not strictly medical, but its comparative-anatomical collection was used regularly in medical teaching. Such collections were institutional – owned by universities (like the collections mentioned in Leiden), hospitals, or medical schools, depending on the local organization of medical teaching. Institutional ownership of collections increasingly replaced private ownership in the nineteenth century, but the shift was by no means absolute.[21] Institutional collections existed before 1800, especially in Leiden, where the university had already built a significant collection in the seventeenth century.[22] Nonetheless, many early modern Leiden professors used their private collections for teaching. The balance shifted after the university acquired the collections of its famous anatomist Bernhard Siegfried Albinus in 1771. In the decades that followed, the Leiden institutional collections expanded rapidly, and university teaching increasingly relied on them. Yet, private collections never completely disappeared.

At least one of Molewater's professors owned a significant personal collection: Jacobus Broers, professor in obstetrics and surgery between 1826 and 1847. During his lifetime, Broers built a collection of pathological preparations that was added to the university hospital's collection after his death.[23] At least two more nineteenth-century professors, both of whom arrived in Leiden after Molewater left, also had private collections. Like Broers, Gerard Suringar, pathology professor from 1843 to 1872, owned a pathological collection. In 1866, Suringar donated over 800 of his preparations to the university.[24] The governors expressed their gratitude with an inscribed silver vase and the preparations were added to the Anatomical Cabinet.[25] Both Suringar and Broers probably stored their preparations at home. Professor Hidde Halbertsma, on the other hand, may have kept his private collection in the Anatomical Cabinet, of which he was the curator. Halbertsma, professor of anatomy and physiology between 1848 and 1865, had built his private collection in the early years of his professorship.[26] Medical teachers elsewhere also had their own collections, either in addition to, or in lieu of, institutional ones.

Molewater himself may also have owned a small collection. The anatomy handbook he used recommended students to gather their own set of preparations. The book was the fourth edition of Georg Hildebrandt's *Handbuch der Anatomie des Menschen* (Handbook of human anatomy, 1830–1832).[27] This edition had been heavily edited by the German anatomist Ernst Heinrich Weber. In his preface, Weber advised students on how best to learn anatomy. One of his recommendations: 'Every student must try to provide himself with the bones of the human body, even if they are to be collected from graveyards.'[28] Not all handbooks were equally explicit about where students should obtain their preparations, but many did recommend that medical students acquire a set of bones – or a complete skeleton, if possible. In his *Handbuch der praktischen Zergliederungskunst* (Handbook of practical anatomy, 1860), the Viennese anatomist Joseph Hyrtl described how, at his university, students 'who do not possess a skeleton' could practise osteology with bones in the dissection hall, implying that at least some students *did* own skeletons.[29] Some handbooks explicitly recommended students to extend their collections with wet preparations.[30]

The advice that students should build their own collections had been circulating for some time: early modern instructors passed it on to their students as well. Londoner William Hunter, one of the leading anatomists in eighteenth-century Europe, lectured: 'I must ... earnestly recommend it to every student, to make and collect as many anatomical preparations as he can.'[31] Hunter recommended his students to acquire a skeleton; several skulls; preparations of the blood vessels, internal organs, and organs of sense and generations; and two preparations of a child's torso, one showing the front of the viscera, the other, the back.

Although we do not know whether Molewater followed the advice to collect some preparations, we do know that he dissected, both in class and in his free time, outside university. For his personal dissections, Molewater relied on animal material. He described his first attempt at dissecting in his diary in April 1835, when he wrote: 'I have received my calf's eye, but unfortunately I lack good scalpels.'[32] Only two days later, he tells us how he and his friend Karel Giltay had successfully dissected a tortoise.[33] Giltay, who at that time had just finished his medical studies, wielded the scalpel quite frequently: only a couple of days beforehand, he had shown Molewater the internal organs of a two-headed goat.[34]

Some students who dissected at home kept preparations after dissecting. Jan Bleuland (1756–1838), for example, who would later become a professor in Utrecht, started his famous comparative-anatomical collection in the first year of his studies in Leiden.[35] Molewater and Giltay might have created and kept preparations as well, but there are no records of this. In general, student collections are not well documented. We do know, however, that stories and images of medical students keeping anatomical objects in their rooms circulated in the nineteenth century. The opening story of this chapter is one example; two satirical drawings by its author, Verhuell, are another. Figure 1.1 is a drawing Verhuell made for the Leiden student almanac, entitled 'Het gevaar van een medicus op kamers te hebben' (The danger of having a medical student in lodgings).[36]

We see a shocked landlady, who has discovered that her tenant is dissecting a human leg under her roof. The landlady enters his room carrying a tray with a bottle that seems to contain alcohol, probably at the student's request – although she might have misinterpreted his reasons. It is surprising that she has not caught the student before, considering the collection of preparations already present on top of his cupboard. Another drawing by Verhuell, entitled 'Een medicus die stil geniet' (A medical student who is quietly enjoying himself), shows a similar image, minus the landlady, with the spotlight on the collection of preparations (Figure 1.2).[37]

Insofar as students did build their own collections, these would have been much smaller than the collections of their institutions and their professors. Although this may not have made them less important to the students, it did make them less visible. A town that hosted medical teaching, like Leiden, housed dozens of collections, and the visibility and accessibility of these collections varied, especially for non-medical audiences. The most visible collections, both to past audiences and present-day historians, were those in museums, and it is to these we now turn.

Figure 1.1 This cartoon, which first appeared in the student almanac, illustrates the danger of having a medical student in lodgings. 'Het gevaar van een medicus op kamers te hebben', drawing by Alexander Verhuell in *Eerste en laatste studentenschetsen* (Arnhem: Gouda Quint, 1882). Courtesy of Leiden University Library, 12568 D 25.

Figure 1.2 A medical student who, according to the caption, is 'quietly enjoying himself'. Many of the preparations seem to be gazing at the student and his work, as is the (still) living dog. 'Een medicus die stil geniet', drawing by Alexander Verhuell in *Zoo zijn er! Studentenschetsen* (Arnhem: Gouda Quint, 1847). Courtesy of Leiden University Library, 12568 D 26.

Museum collections: anatomical preparations on display

In Molewater's day, Leiden University's anatomical museum was housed in an old church, the Faliede Bagijnkerk (Church of the Faille-Mantled Beguines), on Leiden's most famous canal, the Rapenburg. The university's anatomy professors had been using this church since 1594, when Professor Pieter Pauw initiated the building of an anatomical theatre in the church's choir. Back then, the anatomists shared the church with the fencing school, the mathematical school (which offered practical mathematics classes for engineers, in Dutch), and the university library; from 1644, the English Church used the part that had previously been the fencing school. In the following centuries, anatomy would acquire more and more space. In 1725, anatomy professor Bernhard Siegfried Albinus received permission to lecture in the room known as 'the auditorium'. He had to share the room with the mathematical school and the church council, but it was adapted to his needs: a small anatomical theatre was built inside. In 1772, anatomy acquired part of the room used by the English Church. The space was needed to house the collection of preparations the university had bought from Albinus's widow after his death in 1770. By then, the university had been acquiring anatomical objects for over a century and a half; Pieter Pauw had laid the foundation for the university's anatomical collections around 1600. In the seventeenth and eighteenth centuries, preparations were on display in the anatomical theatre (only during summer, when no dissections took place), in the hallway, and above the entrance, where two built-in whale bones could be seen.[38]

The last expansion of the anatomy department before Molewater arrived had taken place between 1819 and 1822. The entire ground floor of the church and a newly built extension were now used by the anatomy department. Figure 1.3 shows the Anatomical Cabinet as Molewater knew it. It had two lecture rooms (both equipped with an anatomical theatre), a dissection hall, a professor's room, a room for the prosector, and a room for the anatomical collections. The latter was the largest, being 22 metres long by 10 metres wide, and 6 metres high. The collections were divided over ten oak-coloured cupboards: one in the middle (13 metres long, 3.25 metres high), eight smaller ones against the long side walls, and one purpose-made for the Albinus collection against the inner wall separating the collections room from the dissection hall.[39] Between the cupboards was a lot of empty space. People could easily walk through the room and observe the preparations, suggesting that the room was used not only for storing the collections, but also for displaying them. Indeed, when preparations were placed on the shelves, they were said to be 'exhibited'.[40]

Many nineteenth-century medical teaching institutions had such spaces where collections were exhibited, often explicitly called 'museums'. Usually, students were one of the intended audiences, or even the only one. Anatomical museums were considered useful instruments in medical teaching until at least the Second World War. Thereafter, they lost much of their prominence, but they never completely disappeared – the present-day anatomical museum at Leiden University is still first and foremost intended as a teaching museum.[41] The nineteenth-century and early twentieth-century use of medical museums in teaching has been best documented for Britain and the United States.[42]

Research on these areas shows that museums had a dual function: they helped students learn anatomical facts while also teaching them how to observe – museums trained the scientific eye.[43] Both goals could also be achieved by other means, of course: students could learn their anatomy by reading books or attending lectures, and atlases trained the scientific eye just as the museum did.[44] But the museum brought several advantages. For one, the knowledge was displayed through preparations, which were considered more 'real' and more attractive to students than drawings. Of course, preparations could be demonstrated in lectures outside the museum as well. Within the museum, however, the preparations could be observed as part of a large ordered system of display; in the lecture hall, this was more difficult, although preparations could become part of a smaller system of display together with, for example, wall charts and drawings on the blackboard.[45] In the first decades of the twentieth century, several Dutch anatomists stressed the importance of their teaching museums, for example in the journal *Methods and Problems of Medical Education*.[46] They explained that the main purpose of the museums was self-study by students. In the

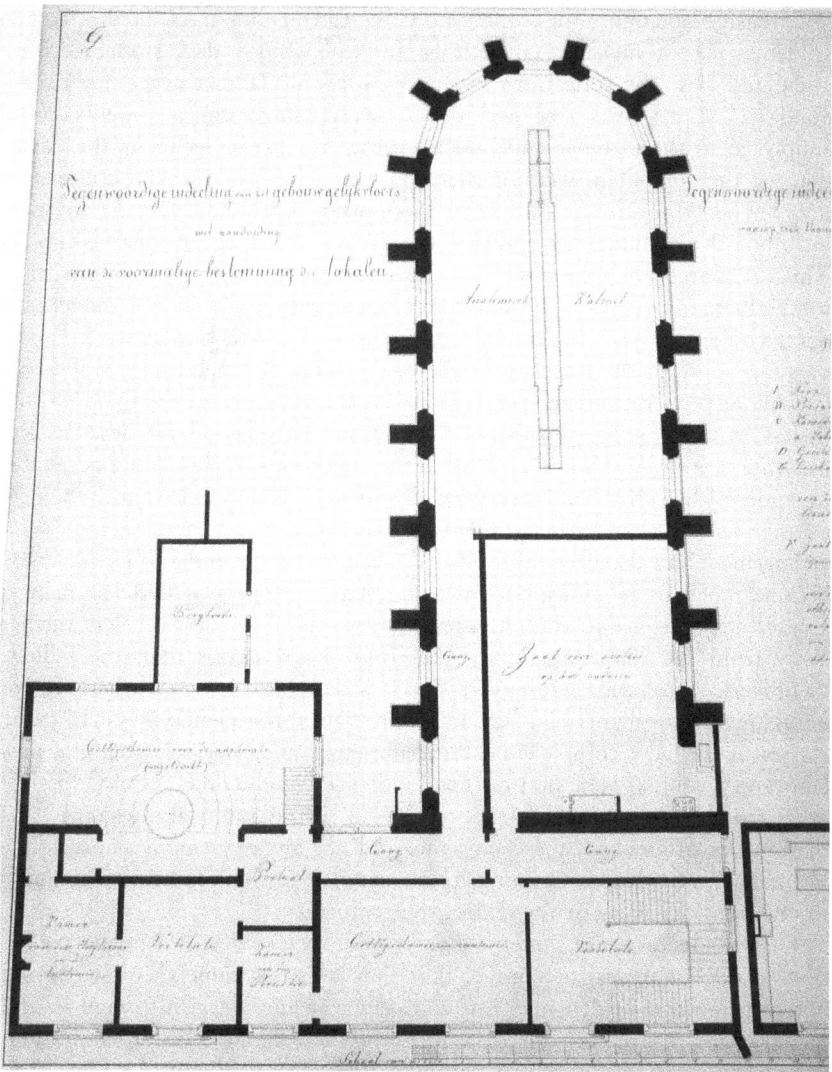

Figure 1.3 Floor plan of the Anatomical Cabinet after the 1819–1822 renovation. The collection room and the dissection hall are in the old church; the remaining rooms are in the newly built extension. Courtesy of Leiden University Library, Archief Universiteitsbibliotheek, 1595–1974 (BA1), inventory number H6, leaf G.

nineteenth century, Dutch anatomists published less on the exact function of museums in medical education, but when we look at actual practices and individual remarks, we see that again, self-study seems to have been the main type of use intended.

Extensive opening hours, announced in student almanacs, facilitated the students' use of museum collections. In Molewater's day, students could come and view the collections in the Anatomical Cabinet every day except Sundays.[47] Entry was free and tickets were unnecessary; students could simply go to the Cabinet and knock on the door to be let in by the custodian. If the custodian was not there, he could be found at his house next door. When the Cabinet moved in 1860, accessibility changed. At the new location, the Cabinet was housed in a newly built educational complex, which it shared with the physics and chemistry laboratories. Museum curator Halbertsma reduced the museum's opening hours after the move: students could now visit the museum four, instead of six, days a week.[48] The opening hours of the dissection hall were extended, suggesting a shift from looking at preparations to creating them – from the eye to the hand. Nonetheless, students still had plenty of opportunities to come and view the collections. Moreover, in another respect, the collections became more accessible to students than they had been: the collections were reorganized, and the new arrangement was tailored to students. As we will see in Chapter 3, the earlier arrangement had also been directed at lay visitors; for them the collections now became much more difficult to interpret. Lay visitors also had to make a greater effort to get to the museum, because the new location was further away from the city centre and other tourist attractions than the Faliede Bagijnkerk had been. Moreover, the new complex had a rather closed, uninviting atmosphere. This was not a problem for students: they had many classes at the new complex, near the museum, and they felt at home in a laboratory context that non-medical audiences might find intimidating. At other Dutch universities, students could also visit anatomical museums.[49] As the century progressed, these museums did not always remain as accessible as the one in Leiden: in Groningen, for instance, opening hours were limited to one hour per week at the end of the century.

That students could, and probably were expected to, visit the collections does not mean, of course, that they actually came. Molewater did not mention the Leiden museum collections in his diary, although he did write about his plans to visit one of the other museums open to university students, the National Museum for Natural History.[50] The Cabinet's annual reports regularly indicate that the collections were used in lectures, but never mention students coming to the museum. And it seems that students were not very fond of the university's museums in general. The student almanac of 1862 states: 'As usual, the various museums were hardly ever visited by students, [but] frequently by strangers.'[51] Here, 'museums' does not necessarily include the anatomical museum – the almanacs usually refer to this museum as a 'cabinet' – but the quotation does reveal students' attitudes towards voluntary museum visits in general, and we have no reason to believe that the anatomical museum differed, in this respect, from the other collecting institutions associated with the university.

Student use of the museum collections was facilitated and was probably expected by professors, but the students were ultimately responsible for their own visits. If they did not wish to visit the museum, they were not forced to. But even students who never visited the museum could not escape anatomical collections. Students had to engage with preparations from the collections in the other teaching spaces as well, although in a different way. In the museum, anatomical collections were displayed; in the lecture room, dissection hall, hospital, and laboratory, collections were handled.

Collections for display, collections for handling

Our hands pose different demands than our eyes, and thus, medical institutions required separate preparations to facilitate both display and handling. Nineteenth-century anatomists carefully distinguished between both types. A mid-nineteenth-century example from Hyrtl's handbook:

> [Preparations], kept in spirits of wine, yet taken out of this for demonstration so as to look at them more carefully, usually do not constitute showpieces in anatomical museums, but are kept ready for use in the side room of laboratories, so-called 'hand museums', which contain all objects that are used often and are subject to a certain change.[52]

In 1909, anatomy professor Jan Willem van Wijhe stated in his opening speech for the new anatomical laboratory at Groningen University that

> in an anatomy laboratory there are at least two collections of preparations; namely, the one exhibited in the museum and the lecture collection. The preparations from the lecture collection are used regularly and are often taken out of the jars, from which their appearance suffers, of course, and so they are not suitable to be exhibited.[53]

The collections for handling went by many names: 'hand museums', 'lecture collections', 'store preparations', 'tank specimens'.[54] These collections differed from display collections in two important respects. First, the preparations had to be robust enough to be handled, and second, the containers used to store the wet preparations had to be easy to open. Both requirements influenced the preparation techniques. Corrosion casts, for instance, rarely appeared in handling collections. To create a corrosion cast, the vessels are first injected and the preparation is then soaked in chemicals that slowly destroy the flesh. The resulting cast highlights the vessel system in great detail, but it is too fragile to be touched. According to a nineteenth-century anecdote, a young man visiting surgeon-anatomist John Hunter's museum in the late eighteenth century was unaware of this and touched a corroded kidney, which immediately broke into pieces. Hunter, who had worked on the preparation for days, perhaps even months, supposedly first demanded to know who the culprit was, but then warned his audience they should not tell him, because 'I shall curse him to the hour of my death'.[55] Although no nineteenth-century technique could produce preparations able to withstand handling for years on end (nowadays, this can be achieved through plastination), many preparation techniques could withstand more handling than corrosion casts. These more robust techniques were preferred for handling collections.[56]

Still, many preparation techniques fitted both display and handling collections. The main technical difference between the two types of collections lay elsewhere: in the containers holding the tissue and the methods used to close them. The aim of closing the container was to slow down the evaporation of the preserving fluid as much as possible. This was best done by sealing the lid as airtight as possible, with materials such as pig's or bullock's bladder, wax, lead, and tin foils. Sealed jars were cumbersome to open – that was the point – but this also made them impractical for handling collections. Handling preparations required different closing methods, as the assistant conservator at the Royal College of Surgeons, Richard Owen, acknowledged to his superiors in a report on the Parisian Muséum d'Anatomie Comparée (Museum of comparative anatomy):

> Every preparation, used at Lecture [at the Muséum] is taken out of the bottle to be demonstrated, and the same thing is done at the request of any scientific visitor who may wish to examine the part in a particular manner; the mode of preventing evaporation is consequently very simple.[57]

Instead of having sealed lids, the bottles in the Muséum were closed with glass lids sunk into a layer of putty that had been applied to the glass – a method Owen considered inferior to the closing techniques used at the College. The bottles themselves were 'of a rude shape, without feet, and mostly of common greenish glass'.[58] When preparations were handled outside of their jars, the appearance of the jars mattered less – they did not even have to be transparent. Handling collections contained easy-to-open glass containers such as stoppered bottles and screw-top jars, but also containers made of non-transparent materials such as metal and wood. In Leiden, large vessel and nerve preparations were stored in wooden containers, although curator Halbertsma would have preferred tight-closing tin tanks to reduce spirit loss and thus keep the costs down.[59] At the Free University in Brussels, large zinc tanks were used.[60] Metal tanks and wooden chests were cheaper and less fragile than glass bottles, and they could be acquired in much larger sizes. Large containers could hold not only large preparations but also multiple preparations together, which reduced the need for expensive preparation fluid.

Multiple preparations were regularly stored together in a single container at the Royal College of Surgeons; almost every page of the storeroom catalogues listed jars containing multiple preparations.[61] In his travel diary (1820), the Dutch medical student Christiaan Tilanus described the storage methods used at the anatomical collection in Heidelberg:

That this collection is constructed not only as a collection, but also to provide a significant number of objects for teaching is proven by the nerve preparations which Mr Tiedemann demonstrated to us. ... All nerves were clearly visible in their mutual relations to their neighbouring parts, blood vessels, muscles, etc.; all of these preparations were stored in a large chest fitted with tin on the inside, with wine-spirit, in which they all were soaked, and even the upper ones could never decay in this habitus.[62]

Tilanus thus distinguished between teaching preparations and collection preparations, but he also suggested that together, these preparations formed one collection. Sometimes handling and display preparations were separated and stored in different spaces; sometimes the different preparations were combined into one collection.

Nowadays, historical handling and display preparations are usually kept together. In museums displaying historical anatomical collections, visitors usually see much more preparations mounted in sealed jars than preparations in stoppered bottles and screw-top jars – and thus, handling preparations seem to have made up a smaller part of the collections than they did. Some of the old handling preparations are (still) not appropriate for display – such as the eight deteriorating fetuses stored together in a jar previously used for Kraft fresh-chilled grapefruit sections, a preparation found in a storeroom at Mount Holyoke College (South Hadley, Massachusetts) by historian Lynn Morgan.[63] Mostly, however, the problem is not that museums do not want to display human remains in recycled pickle jars (their screw-top lids made them excellent for handling preparations), but that handling preparations are less likely to survive over time. All the touching, feeling, squeezing, and poking shortened their lifetimes considerably. Furthermore, because their containers were less airtight than the sealed jars of the display preparations, the fluid evaporated more quickly, making the preparations more prone to decay. Ironically, present-day curators struggle to open the stoppered bottles, which again leads to a shortage of fluid because the preparations cannot be topped up. The preparations have partly dissolved in the fluid, and when the fluid evaporates, the tissue sticks between stopper and bottle, 'gluing' the two tightly together.[64] Back in the nineteenth century, all of these closing techniques – stoppers, screw-tops, and boxes – made it easy to get the preparations out of their containers on a regular basis. Our next question must then be: what happened to the preparations after they were taken out? How, why, when, and where were they handled?

Active observation in the lecture hall

Molewater frequently missed lectures; as we already know, getting up on time was not his strong suit. On 13 October 1834, for instance, he wrote: 'Got up too late to go to Sandifort.'[65] Gerard Sandifort was his anatomy professor. When Molewater skipped his lectures, he not only missed Sandifort's dictation, but also an opportunity to handle preparations. Leiden professors routinely employed preparations as pedagogical tools in their lectures. In a collection report from the 1830s, Sandifort wrote:

> [The collection of the Anatomical Cabinet] is being employed daily in giving lectures both on the anatomy and physiology of Man in healthy and diseased condition, as well as on comparing Man and the Animals, so young students enjoy all its uses.[66]

In the 1850s, comparative anatomy professor Jan van der Hoeven wrote to the university governors that in his lectures he 'constantly used preparations from the Anatomical Cabinet, which then had to be transported to my lecture room at my request'.[67] The obituaries of professors Halbertsma and Zaaijer, who both taught in the second half of the nineteenth century, praised their use of preparations to illustrate their lectures.[68] The photograph in Figure 1.4, dating from around 1890, shows Zaaijer giving a lecture (or at least posing as if he were). He is surrounded by teaching tools including, on the left side of the table, preparations ready to be handled. Behind Zaaijer we see a skeleton, another popular teaching preparation; the 1892 inventory of the Anatomical Cabinet mentions that one of the skeletons was used for teaching and examining students daily.[69]

These examples show that the Cabinet's collections were more than museum collections alone: at least part of them doubled as handling collections. The cupboard containing the collections could be opened easily.[70] When needed for lectures, the preparations were simply taken off the shelves and transported to the lecture rooms down the hall. Other collections also provided lecture preparations. For example, when lecturing at the university hospital, professors used the collection housed in the hospital instead.[71] Sometimes professors presented their own preparations in class, if the Cabinet did not contain what they needed. Later in the century, laboratory collections were also employed. In the early twentieth century, a special elevator was even installed in the pathology laboratory to transport preparations from the laboratory's museum to the lecture room.[72]

In other European cities, lecturers used preparations just as often as their Leiden counterparts. As we have seen above, many medical institutions owned dedicated handling collections for this purpose, sometimes even explicitly called 'lecture collections'. When preparations from these collections were taken to the lecture room, they were usually passed around the audience during the lecture; sometimes they were displayed on a table at the front, so the students could study them afterwards. Students were encouraged to examine the preparations up close. With wet preparations, this was best done by taking them out of their containers. This is obvious in the case of non-transparent containers (such as wooden chests, metal boxes, and earthenware jars), but it also applied to glass jars. Both fluid and glass refracted light and thus distorted the

Figure 1.4 Anatomy professor Teunis Zaaijer in the lecture hall, surrounded by teaching aids – including, on the table, anatomical preparations. Photograph, 1891. Reproduced with permission, Rijksmuseum Boerhaave, Leiden, P07908.

view. Furthermore, in some types of preparations one layer of tissue could obstruct the view on another layer.[73] When a nerve preparation, for example, hangs suspended in a jar, not all the nerves are visible, but once the preparation has been taken out, one can, with one's fingers or a pair of tweezers, pull away the upper nerves to get a good look at the ones below.

By taking the preparations out of their containers, students not only got a closer look, but also could use other senses in addition to sight – in particular touch and smell. Educators encouraged students to use multiple senses, because this supposedly helped them remember what they had learned. Furthermore, by using their senses, students could train and improve them. Nineteenth-century physicians depended on observation to diagnose their patients. It no longer sufficed to ask patients about their symptoms; even university-trained doctors now had to investigate their patients' bodies with their eyes, ears, hands, and noses. Thus, training the senses was an important goal of the practical teaching that became increasingly important as the century progressed.

Students benefited from handling preparations, but the preparations suffered from it. The techniques used for handling collections were meant to withstand touching, but none of them could do so forever; especially when the audience contained students like the ones Scottish anatomist Frederick Knox encountered. In his book on creating and collecting preparations, Knox wrote that 'so far as my own observations go, I am quite certain that few preparations can be entrusted into the hands of students.'[74] Knox therefore understood why some lecturers hesitated to let students handle preparations, but he hoped that his book would help careless students see the errors of their ways. A knowledge of preparation techniques would 'instead of it apparently giving pleasure to many to twist off a toe or finger [of (part of) a skeleton], ... give them real pain from perceiving that they have seriously and permanently injured an anatomical preparation.'[75] And it was not only students who needed educating, in Knox's experience:

> The whole process of preserving anatomical objects in a soft state, requires a care which few persons have the slightest idea of. I remember a gentleman, who had very considerable pretensions to be ranked as a pathologist, presented to the museum [the museum of the Royal College of Surgeons in Edinburgh] a preparation of fracture in the neck of the femur. ... great care was bestowed in dissecting the specimen. By-and-by the gentleman required the preparations for his lectures ... Upon it being returned to me, I found it covered with dust, hairs, &c.; and, on inquiry, found that it had been removed from the jar, handed about the class on a dirt trencher, wiped by the common doorkeeper, with the cloth with which the seats of the classroom had been *carefully* cleaned for the preceding week! I refused to receive the preparation into the collection.[76] (italics in the original)

The Leiden archives make no mention of such unrefined instructors or reckless students, but preparations got damaged nonetheless. This was considered an unavoidable consequence of the standard use of preparations in teaching. As the 1892–1893 report of the Leiden pathological anatomy lab explained, 'from time to time' wet preparations became unusable because they were used in the teaching of almost 200 students. This was no reason

to prohibit students from handling the preparations; the pathological anatomists simply replaced the damaged preparations with new ones, as did
many other nineteenth-century teachers. In Brussels, for example, anatomists kept a list of preparations that had been used up in teaching during
the year, so that the prosectors could replace them.[77] Creating or acquiring
replacement preparations required time and money, and the lecturers' will-
ingness to do this demonstrates how important they considered student
handling of preparation to be.

Students practised anatomy not only by handling preparations, but also by
dissecting corpses. In Leiden, as elsewhere, students were increasingly encouraged and expected to dissect with their own hands. In Molewater's day, the
dissection hall was open to students a few hours a week; in the middle of the
century, anatomy professor Halbertsma extended the opening hours so that
students could practise every day, an opportunity they gratefully seized.[78] But
having more opportunities for dissecting did not make handling preparations in the lecture room redundant. Preparations complemented dissection
in two main ways.

First, preparations could prepare students for dissecting. To dissect effec-
tively, students required at least some knowledge of the human body. (They
also had to get used to working with dead bodies, another area in which
preparations could help, as we will see later in this chapter.) Corpses were
too valuable to allow students to start cutting without any prior knowledge.
They had to have at least some idea of what they were doing, where they
had to cut, and what they were supposed to see inside the body. Opinions
varied as to how much prior knowledge students should have before starting
dissection.[79] Hyrtl advised to commence cutting as soon as possible when
learning general anatomy, but he made sure that he always discussed the
theory *before* students observed parts in the corpse.[80] Students taking
pathological anatomy classes in Leiden had to wait longer: in their first year,
they got a general overview with the help of preparations, and only in the
later years they learned through dissecting.[81]

Second, by handling preparations, students gained knowledge impossible
to acquire through dissecting. Students could dissect only a limited number
of bodies during their studies, which made it unlikely, for example, that they
would observe more than a few pathological conditions during their dissections. Pathological preparations could teach them about all the diseases and
malformations that did not appear on the university dissection tables.
Moreover, preparations could demonstrate bodily properties that the fresh
body did not reveal. Small vessels, for example, cannot be seen with the
naked eye and are thus hard to observe while dissecting. To observe them
properly, one has to make injection preparations, in which the previously
emptied vessels are injected with a fluid foreign to the body.[82] Many different recipes for injection fluid were used in the nineteenth century, but all
had the same purpose: to make even the smallest vessels visible to the naked
eye. The fluid solidified after injection and filled the vessels, enlarging them

and hence facilitating observation. Of course, the small vessels could also be seen through a microscope, but a microscope only shows a small part of the whole at a time, making it hard to get an overview of the relationship and connections between the different vessels. Injection preparations were the preferred method of providing a complete picture.[83] Students also injected vessels during dissection, but this required a considerable amount of time and skill. Therefore, students generally used injection preparations to learn about the build-up of the body's vessels.

Another example of how handling preparations offered knowledge that could not be acquired through dissecting is found in the following quotation from the above-mentioned Frederick Knox. Knox's book gave detailed instructions on how best to preserve different body parts. When discussing the pelvic organs, Knox advised that they should be removed as a whole and then preserved in fluid. He explained that

> preserved in this way, the parts bear a great deal of manipulation … and the catheter may be passed along the urethra, if done with caution, as often as the student has a mind. This manipulation will not, indeed, enable the surgeon to pass the staff or catheter in the living body, but it will give him a vast deal of information which can neither be obtained in practice on the dead nor the living, previous to the removal of the parts from the body.[84]

Students could thus use the preparation to practise procedures such as inserting a catheter. This complemented practising the procedure on a patient or a corpse; as Knox pointed out, handling the preparation provided insights into the make-up of the body that could not otherwise be acquired. Knox did not give further details, but a possible advantage of a preparation over a full body (whether dead or alive) is that practising on a preparation allows one to see what one is doing, instead of having to feel one's way around.

Thus, handling preparations not only prepared students for dissecting, but also provided them with knowledge impossible to acquire through dissecting. During their dissections, students studied preparations alongside bodies in the same hands-on way as they used in the lecture room. And as we will see in the next section, working in the dissection hall required engaging with anatomical preparations in another way as well: through making them.

Making preparations in the dissection hall

In the lecture room students removed preparations from their jars, observed them up close, and felt and smelled them. Such active handling also took place in the dissection hall, to complement the lessons learned from the dissected bodies, but here students engaged with preparations in a more complex way as well: they made them. Dissecting a body and creating a preparation were closely related: student dissection manuals often contained guidelines for making and keeping preparations.[85]

Dissecting sometimes involved a complete body, but more often it meant working with individual body parts. These were easier to distribute among students and store in between sessions. They were also easier to acquire. Bodies were never abundant, as Molewater discovered when he wanted to take the exam to become a surgical doctor in 1851. Molewater had graduated as a *medicinae doctor* in 1840, but when he applied for a job as hospital director, he was asked to become a *chirurgiae doctor* as well, which could be done without having to take additional courses. The exam included a practical session, for which Molewater had to bring his own corpse. Finding one turned out to be difficult, and Molewater wrote to the Leiden surgery professor F. W. Krieger for help. Krieger replied:

> People in Leiden are just as unwilling to go *ad patres* for the benefit of a surgical exam as people in Rotterdam, Americans, &c. are, in other words, we do not have a cadaver available either. ... If you have the opportunity to acquire a cadaver, or part of it, in the meantime [i.e., before the exam], bring it hither; apropos! Might Schneevogt [Gustaaf Eduard Voorhelm Schneevogt, hospital doctor and medical professor in Amsterdam; Molewater befriended him when they studied together in Leiden] not be able to help you? What if you were to write to him that you need a lower limb for your surgical exam, and ask him to send such a *pars cadaveris* to the local anatomy hall before Tuesday?[86]

Krieger proposed that Molewater should ask for a leg if he was unable to find a complete body, which suggests that limbs were easier to come by than entire corpses. This is not surprising; most bodies have two arms and two legs, but only one abdomen. Hence, if a body was used for a demonstration, in all probability at least one arm and one leg would be left over. Furthermore, arms and legs could be taken not only from the dead, but also from the living: amputated limbs probably ended up in the dissection hall. At least, this is implied by the annual lists of available bodies in the Leiden medical faculty: more than once, they referred to both full bodies and some additional limbs that had been acquired.[87]

Students could work on a single body part for weeks on end, slowly turning the fresh flesh into preparations. The time it took to produce a preparation could partly compensate for the shortage of bodies, as is illustrated by the following remark in the Leiden medical faculty's 1850–1851 annual report:

> Concerning the material subsidies for the teaching of anatomy and physiology, Professor Halbertsma remarked that the number of bodies at his disposal for the practical training in anatomy, although anything but high, was sufficient for the meagre number of students that participated. This outcome would, however, not have been possible had not some students concentrated on making delicate vein and nerve preparations, on which, on account of preserving them in spirits, they could work for a reasonably long time.[88]

In between sessions, the unfinished preparations had to be stored so that the tissue neither decayed nor became too rigid to work with. In Leiden, preparations were apparently kept in an alcohol solution. Hyrtl also advised this, but only after one had kept the objects in fresh water as long as possible.[89] Zinc tanks were considered the best containers. These should preferably have ledges that could support a draining grid, because, as Hyrtl put it, 'every practised anatomist knows from experience how awkward and unpleasant it is to transfer still dripping preparations to the dissection table which soon becomes a quagmire; and how much alcohol gets lost in the process.'[90]

In some places, students had to keep a close eye on the body parts they were working with. London student Shephard Taylor, who took anatomy classes in the winter of 1861, wrote in his diary that he 'nearly lost [his] part' because 'an unscrupulous individual' had taken it, since he had failed to attach a card with his name to it.[91] A week earlier, he had not been able to dissect at all, because his body part had been taken away to be used in demonstrations.[92] And a week later, his examination was hampered because the leg he was to be questioned about had 'mysteriously walked off' – another student had accidently taken it.[93]

Once finished, student preparations might be added to the universities' collections. In Vienna, Hyrtl selected the most beautiful ones and displayed them with their makers' names, so as to inspire other students.[94] Pieter Harting, who studied medicine in Utrecht in the 1820s, recalled in his memoirs how some student preparations were added to the university collections, namely the ones on which students had spent an extraordinary amount of time and care.[95]

Students had various reasons to come to the dissection hall and create preparations. In their early years of study, it was a way to learn anatomy and practise their dissection techniques. At the end of their studies, it could form part of their dissertation research; the preparations were then needed to answer particular research questions. In the course of the century, microscopic preparations were increasingly in demand for this. The dissertation written by the Leiden student Johannes Niermeyer offers a typical example of how students worked with microscopic preparations. Niermeyer was a student of Theodorus MacGillavry, professor of pathology between 1877 and 1905. In his neuropathological dissertation, Niermeyer investigated the nerve system of a tetanic rabbit. He referred to other anatomists who had explained the difficulties of discovering tetanus on fresh tissue, and he discussed in detail how he had created microscopic preparations as part of his research.[96]

To make preparations, students needed 'raw material' – preparations required bodies. A key supplier of these bodies was another space that nineteenth-century medical students visited regularly: the academic hospital.

The academic hospital

Medical teaching was founded on bodies. Without bodies, there would be no anatomical demonstrations, student dissections, post-mortems in pathology classes, opportunities to practise surgical procedures, or anatomical teaching collections. Having a large and steady supply of bodies attracted students, but at many European medical institutions teachers struggled to acquire enough bodies to instruct the students already present – just as Molewater had trouble finding a body for his surgical exam. In the Leiden medical faculty's annual report from 1852–1853, the common complaint of a shortage of bodies is followed by a possible solution:

> This year, 14 bodies were available for anatomical demonstrations and the students' practical anatomy training. Although this number was slightly higher than last year, it can still be called small because of the larger demand that existed due to the increased number of students, and it is to be hoped, also for the teaching of anatomy, that the establishment of a city hospital will lead to some improvement.[97]

Apparently, the medical faculty desired more hospital beds because more beds would lead to more deceased patients and hence, the medics hoped, more bodies for teaching. Indeed, hospital patients did sometimes end up on the dissection table, but Dutch anatomists could not simply dissect every deceased patient. In 1869, it was established by law that the family of the deceased had to give their consent, and seeking this consent seems to have been common practice long before this date.[98] The hospital was not the only source of bodies; Leiden University made, or tried to make, arrangements with various prisons as well.[99] Other Dutch universities also relied on both hospitals and prisons for their corpses. Both institutions were important suppliers of medical institutions outside the Netherlands as well, although the specific practices and regulations on the acquisition of bodies for medical teaching varied. In the UK, from 1832 onwards the Anatomy Act stipulated that unclaimed bodies from hospitals, workhouses, and prisons could be dissected before being buried. In practice, this meant that dissected bodies were the bodies of the poor, whose relatives could not pay the funeral costs or were unable to claim the bodies quickly enough.[100] Several Belgian cities had similar regulations with similar consequences.[101]

All in all, some 10 to 20 bodies became available for anatomical dissections in Leiden each year.[102] These bodies had to be used for teaching hundreds of students – when Molewater started his medical studies in 1831, he became one of 129 students at the Leiden medical faculty,[103] and during the nineteenth century, the number of medical students in Leiden increased to between 300 and 400 students.[104] In addition to bodies for dissection, the medical faculty also had access to bodies for autopsies – almost ten times more bodies were available for autopsies than for dissections.[105] During an autopsy, the body was not fully dissected, but cut open only to the extent necessary to establish the cause of death. The medical faculty stressed that, consequently, these bodies could not be used by students to learn anatomy and practise dissecting.[106] In late nineteenth-century Belgium however, this was exactly what happened: historians Tinne Claes and Pieter Huistra have shown that Belgian medical faculties partly resolved the shortage of bodies by increasingly using autopsies, which were (presented as) less controversial than dissections, to transfer anatomical knowledge and skills – for example, by extending the amount of cutting and by letting students handle the knife.[107] In nineteenth-century Leiden, usually the prosector or professor cut open the body during an autopsy, although occasionally students were allowed to as well. If the relatives of the deceased gave their permission, the diseased organ or tissue was removed from the body, after which it was turned into a preparation for the university collection.[108]

Hospital bodies were an important source for anatomical collections. Hospital bodies varied more than prison bodies: the hospital offered both sexes, all ages, and a wide range of pathologies, while prisons supplied mainly adult men. Hence, preparations made from hospital cadavers were more likely to fill a gap in a collection. Thus, when students followed their professor on hospital rounds, they were examining not only patients, but also potential 'raw material' for dissection and preparations. Students realized this, as Molewater's diary shows:

> This morning, I visited the practical classes, including surgery, for the first time [this academic year] and I saw all kinds of miseries. Among other things, a *pièce de caractère* in which Man cut an insignificant figure. A very poor woman lying in bed with two small children, twins, to whom she had recently given birth, both hardly 1 foot long, and whose little cadavers had already been promised to young men by [pathology professor] Broers, in order to be put in spirits. Meanwhile, these moral creatures were still alive, and the mother heard without any sorrow that they would die because she could not provide for them anyway.[109]

The twins were still alive, but Molewater and his fellow students knew they would end up in preparation jars within a few days or weeks at most. While the mother seemingly accepted their deaths 'without any sorrow' and the students who were promised the bodies may have looked forward to it, Molewater struggled with the knowledge. He was neither the first nor the last medical student to have such feelings, but he would need to find a way to handle them if he wanted to become a doctor. Nowadays, for some doctors this includes sharing their emotions with their peers and their patients, but in Molewater's day, and long after, handling emotions mainly meant overcoming them, and learning what has been variously called 'dispassion', 'clinical detachment', 'detached concern', 'medical gaze', and 'necessary inhumanity'.

The last phrase, 'necessary inhumanity', was famously coined by William Hunter in one of his lectures:

> It is dissection alone that can teach us, where we may cut the living body, with freedom and dispatch; and where we may venture, with great circumspection and delicacy; and where we must not, upon any account, attempt it. This informs the head, gives dexterity to the hand, and familiarizes the heart with a sort of necessary inhumanity, the use of cutting instruments upon our fellow-creatures.[110]

Hunter presented dissection as a way of learning the 'necessary inhumanity' that good doctors need to do their job. The dissection hall has often been identified as the place where medical students learn 'dispassion' or 'detached concern' – not just in Hunter's day, but ever since.[111] However, dissection in and of itself is something one needs to be eased into, as Hunter was well aware.[112] In the same lecture, Hunter warned that students should not dissect unprepared, because this 'might even create disgust to a study from which [they] ought to receive pleasure and advantage'.[113] The disgust that might result from dissection was twofold. Cornelis Pruys van der Hoeven, a medical professor at Leiden University between 1824 and 1862, summarized the issue in a metaphor when he wrote about 'studying corpses, to which feelings and smell have to become inured, just as soldiers [have to become inured] to fire and gun smoke'.[114] According to Pruys van der Hoeven, medical students had to train two things: their sense of smell and their feelings. These corresponded to the two forms of disgust that students had to overcome: material and moral disgust. Material disgust is a direct, physical reaction to unpleasant, dirty smells or sights; moral disgust is a struggle with the transgression of social norms.

Primary sources paint the disgusting elements of dissection in all their glory. Our London student Taylor, for example, wrote in November 1861:

> Contrived to remove the intestines from my subject without letting out their contents, an accident that would have won for me the execration of all my fellow-students and perhaps have subjected me to a reprimand from the Demonstrator of Anatomy for my carelessness or want of dexterity in the business.[115]

The contents of a corpse's intestines typically cause material disgust, which is why, in his translation of Ernest Alexandre Lauth's handbook of practical anatomy, the Dutch prosector H. A. Schreuder advised that they should be removed, together with the contents of the bladder, before commencing dissection, so as to reduce grime and stink.[116] Schreuder also suggested removal of hair from the head, beard hair, and pubic hair. These do not provoke a physical reaction, but can cause moral disgust if they make the dissector realize that he (or, from the twentieth century onwards, she) is actually cutting up another human being instead of an 'object'. Taylor again:

Post-mortem examination of a remarkably fine and good-looking girl, who had died of typhoid fever. It made me feel quite sad to see her dead body lying on the post-mortem table, and I could not help but thinking, if she had a lover, how broken-hearted he must have felt at her untimely death.[117]

Hyrtl vividly described how hard it could be to overcome disgust and learn to work on the dead:

The uncommonness of anatomical practices, the ominousness of the surroundings, the emblem of death that impresses every human heart, [these three things] make even insensitive people aware upon their first visit to our rooms with corpses ... that anatomy does not possess an aesthetic side. The first impression it makes on us is cold and serious; there is no cheerful muse greeting us on this gloomy threshold – it is the hand of death which waves us in. How many turn around each year, having looked around in this room [the dissection hall] for the first time, [this room] where only he can feel himself at home whose will has the power, whose dedication has the fervour, whose self-ishness is able to make the sacrifice, which anatomy requires from all of its followers.[118]

How did medical students prepare for the dissection hall, which, as Hyrtl put it, was 'no Eden'?[119] Pruys van der Hoeven explained the practice during his studies in Leiden in the 1810s: 'We, too, started early with human skeletons and bones. This is how we were prepared for the study of corpses.'[120] Handling preparations – bone preparations in this case – helped the students get used to the smell and the emotional impact of working with dead bodies. William Hunter used this solution as well: one of his aims in circulating pre-parations in his lectures was to prepare students for dissection.[121] When doing so, he not only warned the students not to press or bend the preparations in order to avoid their becoming 'injured or destroyed' but also carefully instructed them as to which part of the preparation to examine, which, according to historian Lynda Payne, helped reduce the potential impact of the preparations.[122] For although preparations may seem inherently less disgust-ing than complete, decaying bodies, they not always are. Preparations had to be carefully selected if used to help students overcome their disgust at dissec-tion, otherwise they might very well evoke that same disgust.

The historian of anatomy Erin McLeary has described how many early twentieth-century American students disliked demonstrations of preparations, because, as two of their teachers put it, '[the preparations] are offensive alike to the senses of sight, smell and touch and only the brave or case-hardened person can profit by viewing them.'[123] Besides this material disgust, prepara-tions can also cause moral disgust. This is particularly the case for full-body preparations and preparations of body parts closely connected to human identity, such as the head or, in our time, the brain.[124]

Thus, not all preparations were less disgusting than dissection room corpses, but some of them were; and these preparations helped students ease into dissection, which in turn prepared them for treating patients in a sufficiently detached manner. Whether a preparation evoked disgust depended not only on what it showed, but also on how it had been made. Smelly, materially disgusting preparations could easily be avoided by employing proper techniques. Dry preparations in particular were a safe choice, which is probably why Pruys van der Hoeven's teachers started with bones. Technique also matters when it comes to moral disgust. Medical historian Marieke Hendriksen has shown that eighteenth-century Leiden anatomists tried to make their preparations as skilfully and elegantly as possible, because this was a way to deal with the disgust these objects might otherwise evoke.[125] In this respect, mid-nineteenth-century Leiden professor Teunis Zaaijer resembled his predecessors when he stated in his inaugural lecture that

> Once an anatomist has at his command all means which are offered by technique, something that usually happens only after a lot of practice and effort; and if he is convinced of the necessity of an almost excessive care for purity and pulchritude, [then] an anatomical preparation becomes a painting in his hands, [a painting] on which he depicts the anatomical relations, and *[then] he overcomes the disgust* which anatomy has to evoke if it is practised in another way.[126] (italics mine)

Zaaijer acknowledged that anatomy could easily provoke abhorrence, but he thought that this abhorrence could be overcome with carefully and skilfully created preparations. Such preparations could help students overcome the disgust they felt when first working with dead humans.

Redissecting preparations in the laboratory

Molewater never entered a laboratory as a student, for the medical teaching laboratory was born in the second half of the nineteenth century. In Leiden, the first such laboratory opened in 1865, shortly after the new physical and chemical teaching laboratories (1859), which were almost as important to medical students as the anatomical, pathological, and physiological laboratories. Inside the teaching laboratories, students practised experimental procedures and trained their senses. Preparations were often required for this – for example, to be put under the microscope, an instrument frequently used in the teaching laboratory. Towards the end of their studies, students were increasingly encouraged to do their own research. In this research, they often used preparations – not only recent and freshly made ones (as mentioned above), but also old ones, which they redissected to answer their research questions.

Preparations from the university's handling collections (and possibly also from the students' private collections) were used as empirical material, not only by doctoral students, but also by 'real' researchers, who reinterpreted the preparations as new ideas found their way into medicine. This reinterpretation of older preparations will be discussed in detail in the next chapter.

Hugo Heller was one of the doctoral students who dissected existing preparations. Heller, a student of Zaaijer, wrote his dissertation on *hygroma colli cysticum congenitum*, a malformation of the neck. Having reviewed earlier discussions of the pathology, he turned his attention to an embryo from the Anatomical Cabinet that showed this malformation. He described his examination of the preparation, which involved a redissection:

> After making a cross incision through the skin only ... I loosened the skin with the four flaps so far as was necessary in order to see the boundaries of the tumour ... Under the chin is an opening in the septum, giving access to a hole which was originally covered by membrane ... which in the course of dissection came off together with the skin.[127]

Clearly, little was left of the original preparation in the end, because Heller finally decided to open the chest as well as the tumour by 'splitting the tongue and lower jaw'.[128]

Heller's example is rather extreme, given that he cut up the original preparation in its entirety. While he was by no means the only student to work on preparations from the collections, most tended to leave at least part of the original preparation intact. A survey of all dissertations listed in the most recent (and most complete) bibliographical work on Zaaijer shows that out of the ten named doctoral students, nine worked with preparations.[129] At least six of them used existing preparations in their research. In half of these cases, students used macroscopic preparations to make microscopic ones – which was also a common feature of pathological-anatomical dissertations written under the supervision of Zaaijer's colleagues.[130] Gerardus Couvée was one of the students who transformed part of a macroscopic preparation into a microscopic one. Although he wrote his dissertation in 1900, he worked on a big toe that had been amputated in 1888 and stored in alcohol in the pathology laboratory ever since.[131] The toe contained both a tumour and an interesting pigmentation. To investigate them, Couvée wrote, 'several pieces had been cut off the preparation hardened in alcohol'.[132] Next, he coloured the pieces, after which they were ready for microscopic investigation.

In general, students working on microscopic preparations used fairly recent material, at most one or two decades old. Some students, however, also used much older preparations. For instance, W. Dominicus, Pieter Koning, and Annee Leendert Erkelens researched skulls from the Anatomical Cabinet, including some of the skulls collected by the eighteenth-century anatomists Sebald Justinus Brugmans (Dominicus and Koning) and Bernhard Siegfried Albinus (Erkelens).[133] Although they did not alter the skulls in any way, they re-examined and reinterpreted the skulls using new instruments and ideas, thereby showing once again that preparations were not solely intended to be looked at, but also to be handled.

Conclusion

Nineteenth-century medical students used anatomical collections in all of the teaching spaces they entered. In doing so, they used their hands: they removed preparations from their jars, they observed them with as many senses as possible, they made their own preparations, and they redissected old ones. This helped them learn about the body, master anatomical techniques, answer research questions, and overcome the disgust provoked by dissecting. Anatomical teaching collections have long been neglected in the history of medicine, and when they are discussed, museum collections used in hands-off fashion receive the most attention. Yet, to understand why anatomical collections remained in use throughout the nineteenth century, we need to focus not on museum collections, but on handling collections. These collections fitted seamlessly into the new practical teaching, and students encountered them everywhere, from the lecture hall to the laboratory.

Sometimes, as in Leiden, the two types of collections were partly mixed up and stored together; elsewhere, the two types were kept more separately. At the Royal College of Surgeons in London, for example, only the store collections were handled in teaching; all preparations in the museum collection were meant to stay in the jar when used in lectures.[134]

Although use of the preparations in museum collections was meant to be hands-off, these preparations were sometimes removed from their containers to facilitate observation. But even then, they differed fundamentally from handling collections. In museum collections, the preparations, even if removed from their jars, were observed (either by eye or by hand) *as part of the museum display*. In the museum, the individual preparation, whether inside or outside the jar, gained its meaning from its place in the arrangement of the collection as a whole. Display collections were arranged according to a certain classification on the shelves of the museum, so that students could carefully observe them as part of this order.[135] Handling collections were also arranged according to a certain classification; sometimes even the same classification as that used in the museum. However, when handling collections were used, individual preparations were removed from this classification and transported to another learning space. The arrangement of the collections was not part of the teaching practice in which these collections were used; when the parts of these collections (the individual preparations) were used, they were separated from the whole both spatially and intellectually. The 'whole' was *not* the reason the preparations in handling collections had been collected, unlike the preparations in display collections.

Preparations in handling collections were collected and kept for more prosaic reasons than creating a whole that was more than the sum of its parts. These reasons concerned the practical problems of anatomical research and teaching. Anatomy – whether general, descriptive, pathological, topographical, comparative, microscopic, early modern, nineteenth-century, or present-day – is about bodies. Working with bodies comes with two major practical limitations. First, bodies decay more quickly than they can be dissected. Second, they are scarce and their arrival is unpredictable. Bodies cannot be ordered; at least, not human ones (with a few exceptions, like the Burke and Hare murders in Edinburgh) – animal bodies are often easier to come by, as long as the animal one needs is not too exotic.[136]

Anyone who teaches or researches anatomy has to find a way to overcome decay and secure a regular supply of body parts over time. A common solution is to preserve the material when it becomes available and store it safely for future use. This is still done: in the Leiden 'anatomy skills lab' – as the present-day dissection hall is called – students work with bodies that may be several years old. Preserving tissue (as microscopic slides, as a complete body, or as something in between) is a necessary step in researching and learning about the body; it is the only way to assure the availability of the empirical material when one needs it. When the pieces of preserved material – the preparations – are stored together, a collection is born. Making and collecting anatomical preparations should therefore not only be seen as an end in itself, but also as a means to overcome the limited availability and quick decay of human and (to a somewhat lesser extent) animal bodies – to 'alter time's ... movement', as Harold Cook has put it.[137] With display collections, it is an end; with handling collections, it is a means.

The difference between collecting as a means and collecting as an end, between a focus on the parts and a focus on the whole, is relevant not only for the use of anatomical collections in learning and teaching, but also for the use of these collections in research – to which we now turn.

Notes

1   *Ars*, art, can be interpreted as either 'the art of medicine' or '(the result of) human or technical skill, as opposed to nature'. Unless otherwise noted, all translations in this book are my own.

2   Verhuell, *Zoo zijn er! Studentenschetsen*; Verhuell, *Eerste en laatste studentenschetsen*.

3   Verhuell, 'No. 470, Hoogewoerd'. The most recent reprint I know of is De Wijs, Boven, and Praamstra, *Gruwelkabinet*, 101–8. Other reprints include Verhuell, *Schetsen met de pen*, 96–107; Bervoets, *Nederlandse gruwelverhalen*, 7–13; Zonneveld, *Ontslapen geliefde*, 160–66; Hermans and Zonneveld, *Leiden*, 53–57; Appel and Ross, *Spannende verhalen*, 616–21.

4   Bervoets, 'Alexander Ver Huell', 33.

5   On public fears of medical practices, see for example Richardson, *Death, Dissection and the Destitute*; Richardson, 'Organ Donation'; Stern, 'Dystopian Anxieties'. On gothic horror and the history of anatomy, see for example Marshall, *Murdering to Dissect*; Morton, *Literary Sourcebook*, 82–89.

6 Finger and Law, 'Karl August Weinhold', 163–64; Larson, *Severed*, 239–68.
7 Bervoets, 'Alexander Ver Huell', 34.
8 The diary is now in the city archives in Rotterdam (Dagboekje van J. B. Molewater [Diary of J. B. Molewater], 1833–1835, file 56, Molewater and Rose Family Papers, Stadsarchief, 328, Rotterdam); it has also been published (Molewater, *Studentendagboek*). On Molewater, see also Calkoen, 'Onder studenten', 399–403.
9 Molewater, *Studentendagboek*, 23 (entry 20 August 1833).
10 Knoeff, 'Boerhaave at Leiden', 271–72.
11 On the history of European medical education, see Bonner, *Becoming a Physician*; Buklijas, 'Dissection'; Cunningham, Grell, and Arrizabalaga, *Centres of Medical Excellence?*; Hutton, *Study of Anatomy*; Nutton and Porter, *Medical Education in Britain*. For the United States, see Ludmerer, *Learning to Heal*; Ludmerer, *Time to Heal*; Numbers, *The Education of American Physicians*; Rothstein, *American Medical Schools*.
12 Bonner, *Becoming a Physician*, 33–60.
13 Curtis, 'Swedish in Name Only', 273.
14 Curtis, 'Swedish in Name Only'.
15 'Organiek Besluit Hooger Onderwijs', 2 August 1815 (hereafter cited as RDHE 1815).
16 RDHE 1815, section 'Vijfde titel'.
17 RDHE 1815, articles 169, 177, 180.
18 RDHE 1815, article 43.
19 Alberti, *Morbid Curiosities*; Reinarz, 'Age of Museum Medicine'.
20 Heynsius, 'Inrichting'; Annual report of the Anatomical Cabinet 1884–1885, file 1553, Archief van Curatoren 1878–1953 (hereafter cited as AC3), Leiden University Library.
21 Alberti, 'Owning and Collecting' describes this shift in Britain.
22 On the early modern Leiden collections, see Elshout, *Leidse kabinet*; Hendriksen, *Elegant Anatomy*; Huisman, *The Finger of God*; Knoeff, 'The Visitor's View'.
23 Annual report of the Anatomical Cabinet 1850–1851, file 270, Archief van Curatoren 1815–1877 (hereafter cited as AC2), Leiden University Library.
24 The preparations are described in Suringar, *Pars supellectilis anatomicae*.
25 'Binnenlandsche berigten', *Leydsche Courant*, 11 October 1867.
26 Halbertsma to university governors, 14 March 1852, file 116, document 59, AC2.
27 Molewater mentions the book in his diary: Molewater, *Studentendagboek*, 45 (entry 15 April 1835), 47 (entry 16 April 1835).
28 Hildebrandt, *Handbuch*, 1: xi.
29 Hyrtl, *Handbuch*, 45.
30 Hunter, *Two Introductory Lectures*, 110; South, *Dissector's Manual*, xix.
31 Hunter, *Two Introductory Lectures*, 110.
32 Molewater, *Studentendagboek*, 50 (entry 17 April 1835).
33 Molewater, 52 (entry 19 April 1835).
34 Molewater, 48 (entry 16 April 1835).
35 Knaap, 'Jan Bleuland', 13.
36 Verhuell, *Eerste en laatste studentenschetsen*, Plate 29.
37 First appeared in Verhuell, *Zoo zijn er! Studentenschetsen*.
38 Witkam, 'Anatomieplaats', 22.
39 Witkam, 66–67.
40 Annual report of the Anatomical Cabinet 1856–1857, file 271, AC2. Other annual reports use similar terms.
41 Borgonjen, 'Onderwijs op sterk water', 28. On anatomical museums after 1945, see for example Bozman, 'Modern Medical Museum'; Edwards and Edwards, *Medical Museum Technology*; Hackett, 'The Undergraduate Teaching Medical Museum'.

42   See in particular McLeary, 'Science in a Bottle'; Alberti, *Morbid Curiosities*.

43   Alberti, *Morbid Curiosities*, 164–93; McLeary, 'Science in a Bottle', 19–72.

44   Daston and Galison, 'The Image of Objectivity'.

45   On systems of display, see also Berkowitz, 'Systems of Display'.

46   See for example Barge, 'University of Leiden'; Van den Broek, 'Institute of Anatomy'; Reddingius, 'Het pathologisch-anatomisch laboratorium'; Van Wijhe, *Onderwijs en laboratorium*.

47   Opening hours and access guidelines can be found in student almanacs. See for example Leidsch Studenten Corps, *Studenten-almanak 1839*, 122.

48   Leidsch Studenten Corps, *Studenten-almanak 1860*, 66.

49   For the Utrecht collections see for example Utrechtsch Studenten-Corps, *Studenten almanak 1865*, 33; for Groningen, Groninger Studenten Corps, *Studenten almanak 1885*, 45.

50   Molewater, *Studentendagboek*, 45 (entry 15 April 1835). Comprehensive overviews of museums and collections accessible to students can be found in the student almanacs. The almanac of 1835, for example, lists the University Library, the Museum for Natural History, the Museum of Antiquities, the Cabinet of Plaster Casts and Prints, the Anatomical Cabinet, the Cabinet of Agriculture, the Hortus Botanicus, and the Bibliotheca Thysiana. Leidsch Studenten Corps, *Studenten-almanak 1835*, 78–79.

51   Leidsch Studenten Corps, *Studenten-almanak 1862*, 192.

52   Hyrtl, *Handbuch*, 33–34.

53   Wijhe, *Onderwijs en laboratorium*, 14–15.

54   'Tank' (or 'hand') specimens was used in medical museums in the United States, see McLeary, 'Science in a Bottle', 38n36, 95–96, 196; 'store preparations' (or 'store specimens') was used at the Royal College of Surgeons in London, see for example William Clift, 'Memoranda concerning the sale of old and duplicate specimens of Natural History and Anatomical Articles by the British Museum to the Royal College of Surgeons in London, in the Year 1809', May 1835, document 1/2/2/11, William Clift Papers, Royal College of Surgeons Archives, MS0007, London; and the Annual report of the conservator to museum committee 1876–1877, 2 July 1877, file 8/2/1, Papers of the Hunterian Museum and the Wellcome Museum, Royal College of Surgeons Archives, RCS-MUS, London.

55   Knox, *Anatomist's Instructor*, 7.

56   On techniques particularly suitable for handling collections, see Claes, 'Nobody's Dead', 180.

57   Richard Owen, 'Report to the Board of Curators of the Museum of the Royal College of Surgeons on the Museum d'Anatomie Comparée in the Garden of Plants, Paris', September 1831, file 1/4/1, p. 7, Richard Owen Papers, Royal College of Surgeons Archives, MS0025, London.

58   Owen, p. 7.

59   Annual report of the Anatomical Cabinet 1851–1852, file 270, AC2.

60   Claes, 'Nobody's Dead', 181.

61   See for example Hillman, *Specimens of Invertebrata*; Hillman, *Specimens of Mammalia and Birds*.

62   Deelman and Delprat, *Geneeskunst voor honderd jaren*, 65.

63   Morgan, *Icons of Life*, 1.

64   Martyn Cooke (head of the conservation unit at the Royal College of Surgeons of England), in discussion with the author, 14 April 2011.

65   Molewater, *Studentendagboek*, 44 (entry 13 October 1834).

66   Annual report of the Anatomical Cabinet 1836, file 270, AC2. Teaching use is mentioned almost every year in Sandifort's annual reports on the Cabinet.

67 Van der Hoeven to governors, 6 January 1859, file 127, document 5, AC2.
68 Koster, 'Hidde Justuszn. Halbertsma'; Daniëls, 'Levensbericht van T. Zaaijer', 172; Quant, 'In memoriam T. Zaaijer', 1.
69 Zaaijer, 'Inventaris der verzameling in het Anatomisch Kabinet van de Rijks Universiteit te Leiden', 1892, p. 34, archives Anatomisch Museum (no inventory number), Leiden University Medical Center.
70 Witkam, 'Anatomieplaats', 67.
71 Annual report of the medical faculty 1850–1851, section B.II.b, file 270, AC2; Annual report of the academic hospital 1859–1860, file 271, AC2.
72 Annual report of the pathological-anatomical laboratory 1904–1905, file 1560, AC3.
73 On the difficulties of observing nerve preparations inside their jars see for example Hyrtl, *Handbuch*, 481–82; Lauth, *Nouveau manuel de l'anatomiste*, 688.
74 Knox, *Anatomist's Instructor*, 3.
75 Knox, 3.
76 Knox, 13.
77 Claes, 'Nobody's Dead', 182. On replacing handling preparations, see also for example Hyrtl, *Handbuch*, 34.
78 Leidsch Studenten Corps, *Studenten-almanak 1850*, 108.
79 Buklijas, 'Dissection', 108 describes this for the Habsburg empire.
80 Hyrtl, *Handbuch*, 12.
81 Annual report of the Anatomical Cabinet 1872–1873, file 273, AC2.
82 See Chaplin, 'John Hunter', 118 on eighteenth-century anatomists on this topic.
83 For a Leiden anatomist making this point, see Zaaijer, *Ontleedkundige techniek*, 22–23.
84 Knox, *Anatomist's Instructor*, 124.
85 See for example Hildebrandt, *Handbuch*; see also Hyrtl's summaries of the best-known anatomy handbooks, Hyrtl, *Handbuch*, 46–56.
86 Krieger to Molewater, 13 December 1851, file 51, Molewater and Rose Family Papers, Stadsarchief, 328, Rotterdam.
87 See for example Annual report of the medical faculty (teaching) 1858–1859, file 271, AC2; Annual report of the Anatomical Cabinet 1862–1863, file 271, AC2; Annual report of the Anatomical Cabinet 1876–1877, file 273, AC2.
88 Annual report of the medical faculty 1850–1851, file 270, AC2.
89 Hyrtl, *Handbuch*, 31.
90 Hyrtl, 32.
91 Taylor, *Diary*, 23 (entry 24 January 1861).
92 Taylor, 21 (entry 15 January 1861).
93 Taylor, 24 (entry 1 February 1861).
94 Hyrtl, *Handbuch*, 45.
95 Harting, *Voorheen en thans*, 31.
96 Niermeyer, *Neuropathologische onderzoekingen*.
97 Annual report teaching medical faculty 1852–1853, section B.I.a, file 270, AC2.
98 'Wet tot vaststelling van bepalingen betrekkelijk het begraven van lijken, de begraafplaatsen en de begrafenisregten', 10 April 1869, article 1; Verwaal, 'Sane and Morbid Anatomy', 23–24.
99 See for example Annual report of the Anatomical Cabinet 1862–1863, file 271, AC2; Annual report of the Anatomical Cabinet 1866–1867, file 272, AC2.
100 Richardson, *Death, Dissection and the Destitute*; Hurren, *Dying for Victorian Medicine*.
101 Claes and Huistra, 'Lijken en medische disciplinevorming', 30–31.
102 Annual reports of the Anatomical Cabinet, medical faculty, and university 1851–1878, files 270–73, AC2.

103   Blanken, *Aantal studenten*, 115. The total number of students at Leiden University in 1831 was 742. Although it is difficult to determine nineteenth-century student numbers, Blanken is a reasonably reliable source for Leiden University.
104   Jensma and Vries, *Veranderingen*, 192–93.
105   Annual reports of the Anatomical Cabinet, medical faculty, and university 1851–1878, files 270–73, AC2.
106   Annual report of the university 1856–1857, file 271, AC2.
107   Claes and Huistra, 'Lijken en medische disciplinevorming'.
108   Annual report of the medical faculty 1851–1852, file 270, AC2; Annual report of the university 1856–1857, file 271, AC2. Claes and Huistra describe how in Belgium autopsies were also used to extend the anatomical collections, although there it seems to have been less common to ask permission before removing organs – often, relatives did not even know that the deceased had been autopsied (Claes and Huistra, 'Lijken en medische disciplinevorming').
109   Molewater, *Studentendagboek*, 81 (entry 30 September 1835).
110   Hunter, *Two Introductory Lectures*, 67.
111   See for example Deblon, 'Nieuw beroep', 91; Warner and Edmonson, *Dissection*.
112   Payne, *With Words and Knives*, 111.
113   Hunter, *Two Introductory Lectures*, 108.
114   Pruys van der Hoeven, *Akademieleven*, 16.
115   Taylor, *Diary*, 67 (entry 12 November 1861).
116   The recommendations cannot be found in the original (Lauth, *Nouveau manuel de l'anatomiste*); Schreuder significantly extended Lauth's introductory chapter. Schreuder, *Handboek der practische ontleedkunde*, 1: 18.
117   Taylor, 131 (entry 10 February 1863).
118   Hyrtl, *Handbuch*, 41.
119   Hyrtl, 13.
120   Pruys van der Hoeven, *Akademieleven*, 16.
121   Payne, *With Words and Knives*, 111.
122   Hunter, *Two Introductory Lectures*, 112; Payne, *With Words and Knives*, 111.
123   Robertson and Lundquist, 'Preparation of Pathologic Specimens', cited in McLeary, 'Science in a Bottle', 204.
124   Hoskins, 'Losing and Getting a Head'; Zwijnenberg, 'Brains, Art and the Humanities'.
125   Hendriksen, *Elegant Anatomy*, 108–43.
126   Zaaijer, *Ontleedkundige techniek*, 23.
127   Heller, *Hygroma colli*, 33.
128   Heller, 36.
129   Wallé, *Leiden Medical Professors*, 176–77. As noted in the preface to this volume, the list of doctoral students is by no means complete. This also follows from the fact that during Zaaijer's professorship (1866–1902), at least 455 dissertations were produced in the medical faculty (Van Lieburg, 'Medische promoties', 12. Van Lieburg's numbers end in 1899, three years before Zaaijer's death, meaning that the exact number was slightly higher than 455), and we have no reason to assume that Zaaijer's share was much less than that of the other professors. (In 1866, the medical faculty had seven professors; in 1902, ten.) Yet, although the selection is small, the all-but-one score strongly suggests that a significant number of Zaaijer's doctoral students engaged with anatomical preparations, either by redissecting old ones or by making new ones.
130   See for example two dissertations written under supervision of pathological anatomy professor Daniël Siegenbeek van Heukelom: Haverkorn van Rijsewijk, *Myocarditis tuberculosa*; Josselin de Jong, *Cirrhosis hepatis*.
131   Couvée, *Pigmentatie*.

132 Couvée, 16.
133 Dominicus, *Ontleedkundige aantekeningen*; Erkelens, *Retentio dentium*;
     Koning, *Chineesche schedels*.
134 Richard Owen, 'Report to the Board of Curators of the Museum of the Royal
     College of Surgeons on the Museum d'Anatomie Comparée in the Garden of
     Plants, Paris', September 1831, file 1/4/1, pp. 6–8, Richard Owen Papers, Royal
     College of Surgeons Archives, MS0025, London.
135 See for example McLeary, 'Science in a Bottle', 29.
136 Rosner, *Anatomy Murders*.
137 Cook, 'Time's Bodies', 241.

Bibliography

Manuscript sources

Leiden University Library, Special Collections: Archief van Curatoren 1815–1877;
Archief van Curatoren 1878–1953.
Leiden University Medical Center: Archives Anatomisch Museum.
London, Royal College of Surgeons Archives: MS0007, William Clift Papers;
MS0025, Richard Owen Papers; RCS-MUS, Papers of the Hunterian Museum and
the Wellcome Museum.
Rotterdam, Stadsarchief: 328, Molewater and Rose Family Papers.

Printed sources

Alberti, Samuel J. M. M. 'Owning and Collecting Natural Objects in Nineteenth-Century Britain'. In *From Private to Public: Natural Collections and Museums*, edited by
    Marco Beretta, 141–54. Sagamore Beach: Science History Publications, 2005.
Alberti, Samuel J. M. M. *Morbid Curiosities: Medical Museums in Nineteenth-Century Britain*. Oxford: Oxford University Press, 2011.
Appel, René, and Tomas Ross, eds. *De beste spannende verhalen uit Nederland en
    Vlaanderen*. Amsterdam: Uitgeverij 521, 2007.
Barge, J. A. J. 'University of Leiden Department of Anatomy'. *Methods and Problems of Medical Education* 3(1925): 99–108.
Berkowitz, Carin. 'Systems of Display: The Making of Anatomical Knowledge in
    Enlightenment Britain'. *British Journal for the History of Science* 46(2013): 359–87.
    https://doi.org/10.1017/S0007087412000787.
Bervoets, Jan. 'Alexander Ver Huell en zijn fantastische verhaal "No. 470, Hooge-
    woerd"'. *De Revisor* 9, no. 2(1982): 30–36.
Bervoets, Jan, ed. *Nederlandse gruwelverhalen uit de negentiende eeuw*. Utrecht:
    Veen, 1983.
Blanken, G. H. *Het aantal studenten aan de Hoogeschool te Leiden van 1775 tot en
    met 1868*. Leiden: Steenhoff, 1869.
Bonner, Thomas Neville. *Becoming a Physician: Medical Education in Britain,
    France, Germany, and the United States, 1750–1945*. Oxford: Oxford University
    Press, 1995.
Borgonjen, Renée. 'Onderwijs op sterk water: Het Anatomisch Museum in Leiden
    door ZEE – grafisch en architectonisch ontwerpen'. *de Architect: Interieur* 24
    (2007): 28–29.

Bozman, C. A. 'Planning and Setting out a Modern Medical Museum'. *The Medical Press*, 1958, 842–46.

Broek, A. J. P. van den. 'The Institute of Anatomy, University of Utrecht'. *Methods and Problems of Medical Education* 16(1930): 131–37.

Buklijas, Tatjana. 'Dissection, Discipline and Urban Transformation: Anatomy at the University of Vienna 1845–1915'. PhD diss., University of Cambridge, 2005.

Calkoen, Godert Theodoor Allard. 'Onder studenten: Leidse aanstaande medici en de metamorfose van de geneeskunde in de negentiende eeuw (1838–1888)'. PhD diss., Leiden University, 2012. https://openaccess.leidenuniv.nl/handle/1887/20129.

Chaplin, Simon. 'John Hunter and the "Museum Oeconomy", 1750–1800'. PhD diss., University of London, 2009.

Claes, Tinne. 'Nobody's Dead: The Trajectories of the Corpse in Belgian Anatomy, ca. 1860–1914'. PhD diss., KU Leuven, 2017.

Claes, Tinne, and Pieter Huistra. '"Il importe d'établir une distinction entre la dissection et l'autopsie": Lijken en medische disciplinevorming in laatnegentiende-eeuws België'. *BMGN – Low Countries Historical Review* 131(2016): 26–53. https://doi.org/10.18352/bmgn-lchr.10225.

Cook, Harold J. 'Time's Bodies: Crafting the Preparation and Preservation of Naturalia'. In *Merchants and Marvels: Commerce and Representation of Nature in Early Modern Europe*, edited by Pamela H. Smith and Paula Findlen, 223–47. London: Routledge, 2002.

Couvée, Gerardus. *Pigmentatie der epidermis bij melanosarcoom*. Leiden: Brill, 1900.

Cunningham, Andrew, Ole Peter Grell, and Jon Arrizabalaga, eds. *Centres of Medical Excellence? Medical Travel and Education in Europe, 1500–1789*. The History of Medicine in Context. Farnham: Ashgate, 2010.

Curtis, Stephan. 'Swedish in Name Only: The International Education of Nineteenth-Century Swedish Medical Students and Practitioners'. *History of Science* 50(2012): 257–88. https://doi.org/10.1177/007327531205000302.

Daniëls, C. E. 'Levensbericht van T. Zaaijer'. *Handelingen en mededeelingen van de Maatschappij der Nederlandsche Letterkunde te Leiden*, March 1902(1902–1903): 155–76.

Daston, Lorraine, and Peter Galison. 'The Image of Objectivity'. *Representations* 40 (1992): 81–128. http://doi.org/10.2307/2928741.

Deblon, Veronique. 'Een nieuw beroep en een onverstoorbare ziel: De rol van de anatomie in de gezondheidszorg rond 1800'. In *Vesalius: Het lichaam in beeld*, edited by Geert Vanpaemel, 86–92. Leuven: Davidsfonds, 2014.

Deelman, H. T., and C. C. Delprat, eds. *De geneeskunst voor honderd jaren: Ontleend aan het dagboek-reisjournaal van C. B. Tilanus*. Haarlem: F. Bohn, 1920.

Dominicus, W. *Ontleedkundige aantekeningen betreffende het achterhoofdsbeen*. Leiden: Somerwil, 1878.

Edwards, J. J., and M. J. Edwards. *Medical Museum Technology*. London: Oxford University Press, 1959.

Elshout, Antonie M. *Het Leidse kabinet der anatomie uit de achttiende eeuw: De betekenis van een wetenschappelijke collectie als cultuurhistorisch monument*. Leiden: Universitaire Pers Leiden, 1952.

Erkelens, Annee Leendert. *Retentio dentium*. Leiden: IJdo, 1902.

Finger, Stanley, and Mark B. Law. 'Karl August Weinhold and His "Science" in the Era of Mary Shelley's Frankenstein: Experiments on Electricity and the

Restoration of Life'. *Journal of the History of Medicine and Allied Sciences* 53 (1998): 161–80. https://doi.org/10.1093/jhmas/53.2.161.

Groninger Studenten Corps. *Groninger studenten almanak voor het jaar 1885*. Groningen: Van Boekeren, [1884].

Hackett, Cecil John. 'The Undergraduate Teaching Medical Museum'. In *Proceedings of the First World Conference on Medical Education, London 1953*, 529–37. London: Oxford University Press, 1954.

Harting, Pieter. *Voorheen en thans, 1828–1878: Herinneringen, opmerkingen en wenken door een oud-student*. Utrecht: Greven, 1878.

Haverkorn van Rijsewijk, Karel Theodoor. *Myocarditis tuberculosa*. Rotterdam: De Jong, 1900.

Heller, Hugo. *Over hygroma colli cysticum congenitum*. Leiden: Van Doesburgh, 1881.

Hendriksen, Marieke M. A. *Elegant Anatomy: The Eighteenth-Century Leiden Anatomical Collections*. History of Science and Medicine Library 47. Leiden: Brill, 2015.

Hermans, Tilly, and Peter van Zonneveld, eds. *Leiden: Het land der letteren*. Amsterdam: Meulenhoff, 1985.

Heynsius, A. 'De inrichting van het physiologisch laboratorium'. *Onderzoekingen gedaan in het physiologisch laboratorium der Leidsche Hoogeschool* 1(1869): 1–13.

Hildebrandt, Georg Friedrich. *Handbuch der Anatomie des Menschen*. 4th ed., edited by Ernst Heinrich Weber. 4 vols. Braunschweig: Schulbuchhandlung, 1830–1832.

Hillman, William Augustus. *Catalogue of Specimens of Invertebrata in Store-Bottles (Upper Store-Room, No. 3). Royal College of Surgeons, London*. London: Taylor, 1841.

Hillman, William Augustus. *Catalogue of Specimens of Mammalia and Birds in Store-Bottles (Upper Store-Room, No. 2). Royal College of Surgeons, London*. London: Taylor, 1841.

Hoskins, Janet. 'On Losing and Getting a Head: Warfare, Exchange, and Alliance in a Changing Sumba, 1888–1988'. *American Ethnologist* 16(1989): 419–40. https://doi.org/10.1525/ae.1989.16.3.02a00010.

Huisman, Tim. *The Finger of God: Anatomical Practice in 17th-Century Leiden*. Leiden: Primavera Pers, 2009.

Hunter, William. *Two Introductory Lectures, Delivered by Dr. William Hunter, to His Last Course of Anatomical Lectures, at His Theatre in Windmill-Street*. London: Johnson, 1784.

Hurren, Elizabeth T. *Dying for Victorian Medicine: English Anatomy and Its Trade in the Dead Poor, c. 1834–1929*. Basingstoke: Palgrave Macmillan, 2012.

Hutton, Fiona. *The Study of Anatomy in Britain, 1700–1900*. Body, Gender and Culture 13. London: Pickering & Chatto, 2013.

Hyrtl, Joseph. *Handbuch der praktischen Zergliederungskunst als Anleitung zu den Sectionsübungen und zur ausarbeitung anatomischer Präparate*. Vienna: Braumüller, 1860.

Jensma, Goffe, and H. de Vries. *Veranderingen in het hoger onderwijs in Nederland tussen 1815 en 1940*. Hilversum: Verloren, 1997.

Josselin de Jong, Rodolphe. *Cirrhosis hepatis*. Leiden: IJdo, 1895.

Knaap, Emilie C. van der. 'Prof. dr. Jan Bleuland (1756–1838) en het Museum Bleulandium'. Master's thesis, Utrecht University, [2001].

Knoeff, Rina. 'Boerhaave at Leiden: Communis Europae Praeceptor'. In *Centres of Medical Excellence? Medical Travel and Education in Europe, 1500–1789*, edited by Ole Peter Grell, Andrew Cunningham, and Jon Arrizabalaga, 269–86. The History of Medicine in Context. Farnham: Ashgate, 2010.

Knoeff, Rina. 'The Visitor's View: Early Modern Tourism and the Polyvalence of Anatomical Exhibits'. In *Centres and Cycles of Accumulation in and around the Netherlands*, edited by Lissa Roberts, 155–76. Berlin: Lit Verlag, 2011.

Knox, Frederick. *The Anatomist's Instructor, and Museum Companion: Being Practical Directions for the Formation and Subsequent Management of Anatomical Museums*. Edinburgh: Black, 1836.

Koning, Pieter. *Beschrijving van Chineesche schedels*. Leiden: Van der Hoek, 1877.

Koster, Willem. 'Levensbericht van Hidde Justuszn. Halbertsma'. *Jaarboek van de Koninklijke Akademie van Wetenschappen*, 1866, 38–55.

Larson, Frances. *Severed: A History of Heads Lost and Heads Found*. London: Granta, 2014.

Lauth, Ernest Alexandre. *Nouveau manuel de l'anatomiste*. Paris: F. G. Levrault, 1829.

Leidsch Studenten Corps. *Studenten-almanak voor het jaar 1835*. Leiden: Herdingh, [1834].

Leidsch Studenten Corps. *Studenten-almanak voor het jaar 1839*. Leiden: Herdingh, [1838].

Leidsch Studenten Corps. *Studenten-almanak voor het jaar 1850*. Leiden: Gerhard, [1849].

Leidsch Studenten Corps. *Leidsche studenten-almanak voor 1860*. Leiden: Engels, [1859].

Leidsch Studenten Corps. *Leidsche studenten-almanak voor 1862*. Leiden: Engels, [1861].

Lieburg, Mart J. van. 'De medische promoties aan de Nederlandse universiteiten (1815–1899)'. *Batavia academica: Bulletin van de Nederlandse Werkgroep Universiteitsgeschiedenis* 5(1987): 1–17.

Ludmerer, Kenneth M. *Learning to Heal: The Development of American Medical Education*. Baltimore: Johns Hopkins University Press, 1996. First published 1985 by Basic Books.

Ludmerer, Kenneth M. *Time to Heal: American Medical Education from the Turn of the Century to the Era of Managed Care*. Oxford: Oxford University Press, 2005.

Marshall, Tim. *Murdering to Dissect: Grave-Robbing, 'Frankenstein' and the Anatomy Literature*. Manchester: Manchester University Press, 1995.

McLeary, Erin Hunter. 'Science in a Bottle: The Medical Museum in North America, 1860–1940'. PhD diss., University of Pennsylvania, 2001.

Molewater, Jan Bastiaan. *'Hoe zal het met mij afloopen': Het studentendagboek 1833–1835 van Jan Bastiaan Molewater*. Edited by Henk Eijssens. Hilversum: Verloren, 1999.

Morgan, Lynn M. *Icons of Life: A Cultural History of Human Embryos*. Berkeley: University of California Press, 2009.

Morton, Timothy, ed. *A Routledge Literary Sourcebook on Mary Shelley's 'Frankenstein'*. London: Routledge, 2002.

Niermeyer, Joh. Hendr. Ant. *Neuropathologische onderzoekingen*. Leiden, 1879.

Numbers, Ronald L. *The Education of American Physicians: Historical Essays*. Berkeley: University of California Press, 1980.

Nutton, Vivian, and Roy Porter, eds. *The History of Medical Education in Britain*. Amsterdam: Rodopi, 1995.

Payne, Lynda. *With Words and Knives: Learning Medical Dispassion in Early Modern England*. The History of Medicine in Context. Aldershot: Ashgate, 2007.

Pruys van der Hoeven, Cornelis. *Akademieleven*. Amsterdam: Van der Post, 1866.

Quant, C. A. J. 'In memoriam T. Zaaijer'. *Studenten Weekblad Minerva* 27, no. 29 (1903): 1.

Reddingius, R. A. 'Het pathologisch-anatomisch laboratorium'. In *Academia Groningana, MDCXIV–MCMXIV: Gedenkboek ter gelegenheid van het derde eeuwfeest der Universiteit te Groningen*, edited by Is. van Dijk, J. W. Moll, G. C. Nijhoff, J. Simon van der Aa, J. Huizinga, J. H. Kern, and H. J. Hamburger, 500–511. Groningen: Noordhoff, 1914.

Reinarz, Jonathan. 'The Age of Museum Medicine: The Rise and Fall of the Medical Museum at Birmingham's School of Medicine'. *Social History of Medicine* 18 (2005): 419–37. https://doi.org/10.1093/shm/hki050.

Richardson, Ruth. *Death, Dissection and the Destitute: The Politics of the Corpse in Pre-Victorian Britain*. 2nd ed. Chicago: University of Chicago Press, 2000.

Richardson, Ruth. 'Human Dissection and Organ Donation: A Historical and Social Background'. *Mortality* 11(2006): 151–65. https://doi.org/10.1080/13576270600615419.

Robertson, H. E., and Richard Lundquist. 'The Preparation of Pathologic Specimens for Exhibition Purposes'. *Bulletin of the International Association of Medical Museums* 13(1934): 31–32.

Rosner, Lisa. *The Anatomy Murders: Being the True and Spectacular History of Edinburgh's Notorious Burke and Hare and of the Man of Science Who Abetted Them in the Commission of Their Most Heinous Crimes*. Philadelphia: University of Pennsylvania Press, 2010.

Rothstein, William G. *American Medical Schools and the Practice of Medicine: A History*. Oxford: Oxford University Press, 1987.

Schreuder, H. A., trans. *Handboek der practische ontleedkunde*, by Ernest Alexandre Lauth. 2 vols. Leiden: Van den Heuvel, 1839.

South, John. *The Dissector's Manual: A New Edition, with Additions and Alterations*. London: E. Cox, 1825.

Stern, Megan. 'Dystopian Anxieties versus Utopian Ideals: Medicine from Frankenstein to the Visible Human Project and Body Worlds'. *Science as Culture* 15(2006): 61–84. https://doi.org/10.1080/09505430500529748.

Suringar, Gerard Conrad Bernard. *Pars supellectilis anatomicae, sive Catalogus speciminum pathologico-anatomicorum, quae in usus privatos a se collecta, praeparata et ordine disposita, Academiae Leidensi vivus*. Leiden: Drabbe, 1866.

Taylor, Shephard. *The Diary of a Medical Student during the Mid-Victorian Period, 1860–1864*. Norwich: Jarrold, 1927.

Utrechtsch Studenten-Corps. *Utrechtse studenten almanak voor het jaar 1865*. Utrecht: Van Schoonhoven, [1864].

Verhuell, Alexander. 'No. 470, Hoogewoerd: Eene legende'. In Leidsch Studenten Corps, *Studenten-almanak voor het jaar 1847*, 152–60. Leiden: Gebhard, 1847.

Verhuell, Alexander. *Zoo zijn er! Studentenschetsen*. Arnhem: Gouda Quint, 1847.

Verhuell, Alexander. *Schetsen met de pen*. Amsterdam: Gebhard, 1853.

Verhuell, Alexander. *Eerste en laatste studentenschetsen*. Arnhem: Gouda Quint, 1882.

Verwaal, Ruben. 'Sane and Morbid Anatomy: The Studies and Specimens of Professor Jan Bleuland'. Master's thesis, Utrecht University, 2010.

Wallé, Dalila. *Leiden Medical Professors 1575–1940*. Leiden: Museum Boerhaave / Leids Universitair Medisch Centrum, 2007.

Warner, John Harley, and James M. Edmonson. *Dissection: Photographs of a Rite of Passage in American Medicine, 1880–1930*. New York: Blast Books, 2009.

Wijhe, Jan Willem van. *Onderwijs en laboratorium bij de ontleedkunde*. Groningen: Wolters, 1909.

Wijs, Ivo de, Erica van Boven, and Olf Praamstra, eds. *Het gruwelkabinet: Dertien horrorverhalen uit de negentiende eeuw*. Amsterdam: Athenaeum-Polak & Van Gennep, 2010.

Witkam, H. J. 'Over de anatomieplaats, de Albinussen en de Sandiforts'. Typescript, 1968.

Zaaijer, Teunis. *Het gewigt eener doelmatige ontleedkundige techniek*. Leiden: Hazenberg, 1866.

Zonneveld, Peter van. *De ontslapen geliefde: Verhalen uit de Romantiek*. Amsterdam: Bakker, 1983.

Zwijnenberg, Robert. 'Brains, Art and the Humanities'. In *Neurocultures: Glimpses into an Expanding Universe*, edited by Francisco Ortega and Fernando Vidal, 293–310. Frankfurt am Main: Peter Lang, 2011.

2 Make do and mend

How researchers used old collections in new medicine

18 July 1819. Dusk. Leiden professor Sebald Justinus Brugmans had been working all day in the botanical garden and the natural history cabinet. Suddenly, his chest hurt and his stomach cramped. Initially, a simple bloodletting seemed to solve the problem. But the stomach cramps returned, and soon grew worse. Gastroenteritis, followed by gangrene. Four days after he had felt the first pain, the professor died.[1] He left behind a collection of roughly 4,000 anatomical preparations.

Brugmans's death did not spell the end for his collection. Far from it: his preparations were not used in published research until after he had died. Brugmans used his preparations primarily during his classes; just like his contemporaries, he valued teaching more than research.[2] Research was done, of course, but professors often communicated their findings solely through teaching – 'publish or perish' was a phrase yet to be coined. The medical historian Antonie Luyendijk-Elshout extensively studied the history of the eighteenth-century Leiden anatomical collections, but she did not find a single publication in which Brugmans mentioned his collection. From this she concluded: 'To Brugmans, these preparations have probably seldom served for detailed study.'[3] Although this may have been true of Brugmans, it was certainly not the case for his successors. As this chapter will show, researchers in nineteenth-century Leiden regularly used the Brugmans collection in their publications. They also used the collections of Johannes Rau, Bernhard Siegfried Albinus, and Andreas Bonn – all anatomists who lived and worked decades before the researchers discussed in this chapter.

Nineteenth-century Leiden researchers relied heavily on the old, mostly eighteenth-century collections. In 1850, the Anatomical Cabinet housed approximately 8,000 preparations, of which around 7,500 had been created before 1815.[4] New preparations were added, but the majority of these came from estates, meaning that even many 'new' acquisitions had been made by anatomists from earlier generations. Some researchers owned private collections, but they usually built these with an eye to teaching, not research, as teaching was the main source of income for most researchers. Furthermore, the private collections were small compared to the university collections.

Thus, on the whole, nineteenth-century Leiden researchers had to make do with preparations created by their predecessors.

In the nineteenth century, reuse of preparations made by earlier anatomists was widespread. Nineteenth-century universities, hospitals, and medical schools regularly built their collections around former private collections, which had often been created in the eighteenth century.[5] In Utrecht, researchers used the preparations of Jan Bleuland (1756–1838); in Groningen, researchers relied on the collection of Petrus Camper (1722–1789). Some nineteenth-century collections contained more newly produced preparations than those in Leiden. Whereas in Leiden, according to the annual reports, usually fewer than ten freshly made preparations were added to the collections each year, at, for example, the Royal College of Surgeons in London, thousands of preparations were produced in-house during the nineteenth century. Yet, even at the College, researchers continued to use older preparations alongside newer ones, as we shall see from the examples later in this chapter.

This may seem surprising, as nineteenth-century medical research differed profoundly from its eighteenth-century predecessor. New disciplines had emerged, such as comparative anatomy, pathological anatomy, and developmental embryology.[6] Furthermore, the old disciplines of anatomy and physiology had been transformed completely.[7] The emerging and changing disciplines used different spaces, such as the laboratory and the clinic; different methods, such as microscopy; and different concepts, such as the cell.[8] All of these changes reached Leiden, although often later than they reached many other places.[9] None of the changes obviated the need for collections, but all of them made new demands of them. And yet, old collections continued to be used in the new medicine.

It seems that the same preparations could be used in medical research practices for a long time, even as these practices changed profoundly. In this chapter, I investigate the nineteenth-century afterlife of the Brugmans collection to understand how such prolonged use was (and still is) possible. First, I will explain how the material characteristics of preparations in general make them flexible. Subsequently, I will sketch the background to the Brugmans collection and explain how it ended up in the Leiden Anatomical Cabinet. I will then analyse how nineteenth-century researchers (re)used Brugmans's preparations in three fields of study: physical anthropology, pathological anatomy, and comparative anatomy.

Preparations: made of what they represent

Preparations can display facts about the human (or animal) body. They can show how syphilis damages the skull, what the inside of the heart looks like, and how muscles are connected to bones. One can also write such facts down, but anatomists may prefer to communicate them with the aid of preparations, not only because these can show more than words can tell, but also because preparations serve as evidence to back up anatomical claims.

Preparations, then, are end products in the making of knowledge. They are a stabilized version of a natural phenomenon (namely, a living body) that helps demonstrate, explain, and communicate facts. The German philosopher of biology Hans-Jörg Rheinberger has called such stabilized objects 'epistemologica'.[10] Other examples of epistemologica include anatomical models, meteorological graphs, and geological maps.

But preparations can do more than just demonstrate existing facts: they can also produce new ones. As we will see below, researchers, like students, actively handled preparations. In doing so, researchers reinterpreted and even redissected older preparations. For them, preparations were not just end products displaying existing facts, but also raw materials that could be used to produce new knowledge. Preparations thus played a dual role in research: they were both finished and unfinished; both representations of ready-made knowledge and raw material for new facts; both, if you want, artefacts and naturalia. This distinguishes preparations from most other epistemologica, which can only play a single role: that of exposing existing facts. To understand where this difference comes from, let us compare anatomical preparations with anatomical models.

Models and preparations are similar in that they both *represent* a particular object of inquiry. Yet they are also fundamentally different, because preparations are a very peculiar kind of representation. Rheinberger argues that 'normal' representations are defined by two characteristics.[11] The first is a change to a different medium: an anatomical model is made of wax, papier-mâché, or plastic, while its object consists of human tissue. The second is a rule (or set of rules) that maps the object to the medium. Preparations are atypical because they lack the first characteristic. They are representations and yet they are (at least partly) made out of the same material as their objects. Unlike a model, a kidney preparation does not consist of, for example, wax or papier-mâché – it consists of *kidney*. Preparations are *made of what they represent*, which enables them to play their second role: that of an unfinished product, empirical material, used to answer questions other than the ones they were made to answer.

Rheinberger is referring to this capacity when he writes: 'The essence of organic preparations qua knowledge objects resides in this material complicity [being made of what they represent], which ensures their duration and the permanent possibility of their epistemic recall.'[12] Rheinberger's observation is crucial because it pinpoints *why* old preparations could (and still can) be reused again and again. This may not immediately be clear because, unfortunately, the observation is also rather dense. To unravel it, consider what it takes for seemingly finished made-objects to be reused in producing new knowledge, as happened to the Brugmans preparations in the nineteenth century. Above all, they need to facilitate reinterpretations. Both preparations and models are created with certain questions, or at least vague ideas, in mind. Their makers create them to generate or prove new knowledge relating to these questions or ideas or, in the case of preparations and models intended solely for teaching, to demonstrate known facts. But as time goes by, (new) researchers start working with different questions and different ideas. For example, instead of aiming to describe a tumour macroscopically, they want to understand it on a cellular level. To answer the new questions, they need to either make new preparations and models, or reinterpret the old ones. Sometimes, reinterpreting means simply writing a new label – when renaming a species or reclassifying a plant, for example. But often, reinterpretations are more complex and require new empirical data: additional information not directly offered by the object. Take the example of the tumour: a cell theory-related reinterpretation requires the tumour's microscopic structure, but neither a macroscopic preparation nor a macroscopic model represents this structure.

When it comes to such complex reinterpretations, preparations have an advantage over models: they are more likely to contain the required information because they are made of what they represent. Both models and preparations contain information, and both may contain more information than strictly required for their original purpose. But models only contain the information *added* by the modeller, whereas preparations contain all information *not taken away* by the prosector. Therefore, models only contain information that was accessible to their maker. For example, nineteenth-century papier-mâché models of snails *never* contain the snail's DNA structure, because the molecular level could not be accessed by the dissecting and model-making instruments of the time. A nineteenth-century alcohol preparation of the same snail, on the other hand, *does* contain its DNA structure. The preparation maker did not have access to it, but he (or perhaps occasionally she) did not need to: the structure was nevertheless included in the material. Therefore, with the preparation, it is possible to 'go back' to the 'original' object of inquiry (the snail) and extract the DNA structure at a later moment in time. Similarly, a macroscopic tumour preparation can be sliced and examined microscopically to study its cells, which is impossible with a tumour model. None of this is to say that models can never be reinterpreted, but their reinterpretation is much more difficult.

The 'going back' to the object of inquiry is what Rheinberger calls 'epistemic recall'. Rheinberger argues for this epistemic recall in theory; with the example of the Brugmans collection, I will show how it worked in practice. For it was because of the continuous reinterpretation of preparations that Brugmans's collection remained useful for medical research throughout the nineteenth century.

Brugmans and his collection

Sebald Justinus Brugmans (1763–1819, Figure 2.1) collected his first naturalia in his parents' backyard, where he searched for shells and stones as a child.[13] He continued building collections for the rest of his life. When he studied in Groningen, he collected stones in areas surrounding the city; this collection formed the empirical foundation of his first doctoral dissertation, in philosophy, which he completed in 1781.[14] For his second doctorate, in medicine, he spent several years studying in Leiden. During that period, he assisted Leiden professor Dionysius van de Wijnpersse in arranging the natural history collection of the deceased medical professor Wouter van Doeveren (1730–1783).[15] Brugmans received his medical degree in 1785 from the University of Groningen. Soon afterwards, he was appointed professor in Franeker. He left a few months later, after having been offered a position in Leiden. The Leiden governors appointed him as a professor in the philosophical faculty, where he taught courses on botany, mineralogy, and zoology. But Brugmans was not satisfied with this position and longed for a professorship in the medical faculty. After some lobbying, he succeeded in 1791. This displeased the other medical professors, who feared they would lose students, and hence money, to Brugmans.[16] Brugmans's teaching was widely praised; he was said to speak appealingly and without notes. To illustrate his lectures, he built a collection of anatomical preparations – the same collection we will follow in this chapter.[17]

Brugmans remained a professor in Leiden until his death in 1819, but he regularly took on activities outside the university as well, which extended the possible sources for his collection. He advised subsequent governments of very different political leanings on health issues. He led the national Military Medical Services for 20 years, advised on cattle plague, and contributed to a national pharmacopoeia, the *Pharmacopoea Batava*.[18] His work on battlefields and in military hospitals offered him ample opportunity to collect pathologies and foreign skulls. His research on the cattle plague required animal dissections, during which he may have created preparations; another possible source for his animal preparations was the animals kept in the university's botanical garden.[19] Furthermore, several of Brugmans's connections – including the French comparative anatomist Georges Cuvier, but also his subordinates in the Military Medical Services – sent him skulls, bones, fossils, and other objects, sometimes fully prepared.[20]

In 1817, Brugmans offered his collection to the university, 'on the reasonable condition of compensation'.[21] The immediate cause of Brugmans's offer – and of the university governors' acceptance – was the 1815 Royal Decree on Higher Education, which obliged all universities to own several types of anatomical preparations.[22] This included comparative anatomy preparations, which were lacking in the Leiden University collections, but well represented in Brugmans's collection. Around half (2,093) of Brugmans's 4,081 preparations were considered comparative anatomical; just over a quarter (1,154) were pathological; the remaining ones were mainly natural history objects (635) and fossils (141).[23] The large number of comparative anatomy preparations made the governors keen to acquire the collection. They agreed with Brugmans upon a 'compensation' of 30,000 guilders to be paid in six annual instalments.[24] However, Brugmans died two years into the agreement, which prompted his widow to reopen the negotiations. She secured an additional 4,000 guilders for herself, because of

C. H. Hodges. pinx.   W. van Senus. sculp.

SEBALD JUSTINUS BRUGMANS.

Amsterdam 1829. by W. van Senus. Prinsengracht, by de Spiegelstraat N.127.

Figure 2.1 Portrait of Sebald Justinus Brugmans. Line engraving by W. van Senus, 1829, after C. H. Hodges. Courtesy of the Wellcome Collection, London. (CC BY 4.0)

the new preparations made by Brugmans that had not been included in the first deal, and because she also offered the collection cupboards to the university.[25] From November 1819, the university officially owned Brugmans's collection.

The governors appointed Gerard Sandifort, curator of the Anatomical Cabinet, as the supervisor of the Brugmans collection, and asked him to catalogue it.[26] Sandifort responded with caution: he admitted that a catalogue would increase the collection's value, but explained that cataloguing would be difficult and time-consuming.[27] He was willing to invest the necessary time, but asked for two things in return. First, he wanted to keep teaching the comparative anatomy classes; second, he wanted the comparative-anatomical part of the collection to be housed in the Anatomical Cabinet. It went without saying that the general and pathological anatomy preparations would be added to the Cabinet, but the preparations of comparative anatomy would have been useful in the university's natural history cabinet as well. Sandifort acknowledged this, but he claimed that they were better suited to the Anatomical Cabinet because of their ultimate aim of illustrating the structure and functions of the human body. When on 30 September 1820 the governors ultimately decided on the fate of the comparative anatomy preparations in the Brugmans collection, Sandifort had already finished half of the catalogue, which consisted of descriptions based on Brugmans's labels and Sandifort's own investigations.[28] The governors allowed him to keep all the comparative anatomy preparations, as he had wanted, but decided that the natural history preparations would be housed in the new National Museum for Natural History, into which the university's natural history cabinet had been incorporated.

Obviously, the new museum collected natural history objects, but what exactly were these? And how did they differ from comparative anatomy objects? A letter from Sandifort sheds light on these questions. On 21 October 1820, he wrote to inform the governors about which Brugmans preparations he intended to transport to the Museum for Natural History:

> Since Your Highly-Learned Honourables have demanded that all objects that do not directly belong to the collection of comparative anatomy, but are more related to natural history, should be added to the Cabinet of Natural History, I shall not fail to deliver to this Cabinet all objects kept in liquor, including the collection of shellfish, as instructive as extensive, &c.; the dried or stuffed animals; all fossil bones; and, further, one specimen of every skeleton and animal head that we have in duplicate; I hope this meets your intentions.[29]

To Sandifort, natural history objects were whole-body preparations of animals (either stuffed, dried, or in fluid) and animal bones, skeletons, and fossils. Sandifort's definition matches that found in a Ministerial Decree issued two months later, on what the Museum for Natural History should and should not collect:

> 2. In this museum, animal species (with the exception of Man) and their complete or partial skeletons will be brought together and kept, as well as fossils and minerals.

> 3. No preparations of the individual animal organs, either pathological or physiological, fall within the scope of this Cabinet.[30]

The museum was allowed to collect complete animals, animal skeletons, fossils, and minerals; these were considered to fall under the heading of natural history. Preparations of animal organs, however, were not added to the museum, as these were seen as part of comparative anatomy, not natural history. Their proper home was the Anatomical Cabinet, at least until around 1860. At this point, curator Hidde Halbertsma used the Cabinet's move to rearrange and reclassify the preparations, and to get rid of the preparations he deemed irrelevant for medical research and teaching. Among other things, he disposed of part – but not all – of Brugmans's comparative anatomy preparations. These were moved to the Museum for Natural History, the scope of whose collections had been officially extended in 1859.[31]

These days, the collection is spread across various institutions. Three Leiden museums house most of the remaining preparations: the university's Anatomical Museum, Naturalis (the successor to the National Museum for Natural History), and Museum Boerhaave, a museum devoted to the history of science and medicine.[32] As we have seen, the segmentation of the collection began soon after its acquisition. Historian Hans de Jonge has condemned the governors' decisions:

> Due to mismanagement by the Leiden university governors, who had no idea what kind of collection they had acquired, the collection was broken up as early as 1820 ... The governors made the tragic decision to divide the Brugmans collection between both institutions [the Museum for Natural History and the Anatomical Cabinet] ... The governors did not understand that the division completely negated the fundamental principle of the collection, the comparison of skeletons and organ systems throughout the animal species right up to humankind.[33]

De Jonge implies that the governors should have preserved the collection in accordance with Brugmans's 'original' intentions, and interprets their failure to do so as born of ignorance. However, De Jonge does not allow for the fact that the governors did not acquire the Brugmans collection because they wanted to preserve material heritage, but because they believed the professors could use the preparations for teaching and research. (That said, the governors were keen on using the anatomical collections, including Brugmans's, as status symbols, because of their connection to the past, as we will see in Chapter 4.) The professors could, indeed, use them for this purpose, but their ideas about research and teaching differed from Brugmans's, and it was therefore necessary to reinterpret the collection. Splitting up the collection formed part of this reinterpretation, and as such, it reflects not a lack of insight, but changing ideas on research and teaching. Brugmans's preparations were flexible enough to be adapted to these changing ideas, mainly because they – like all preparations – were made of what they represented. In the following sections, I will discuss the reuse of Brugmans's preparations in three medical fields: physical anthropology, pathological anatomy, and comparative anatomy. We will see how researchers extracted new information from the preparations, and how the preparations continued to remain relevant in medical research throughout the nineteenth century.

Physical anthropology

Physical anthropology has been defined as the study of the similarities and differences between the bodies of groups of people.[34] It focuses mainly on differences in the structure of the body. To establish these differences, researchers measured and compared either the bones of the dead or the bodies of the living. The former is called craniology or craniometry; the latter, anthropometry. In the nineteenth century, both fields were tied to medicine. Their practitioners were usually trained as medical doctors and published in medical journals, and the collections they required were, at least in the early days, housed in medical institutions. Here I focus on craniometry, because this approach relied heavily on anatomical collections.

In the Netherlands, craniometry became a well-defined area of study in the mid-nineteenth century. Until 1900, Leiden was the main centre of the field, with initially the Anatomical Cabinet and then, from 1880 onwards, the Ethnographical Museum as the leading institution.[35] Leiden professors Teunis Zaaijer and Jan van der Hoeven were among the first practitioners. They were 'armchair anatomist-anthropologists'.[36] They did not go out into the field, but relied completely on the skulls and bones already present in their local collections. They used whatever material came to them – either from overseas or from the past. In the early days in particular, they relied on older preparations; the Anatomical Cabinet received very few new anthropological preparations between 1835 and 1860.[37] Among these older preparations were the anthropological objects from the Brugmans collection.

As mentioned above, Brugmans collected foreign skulls on battlefields. He also received skulls (and other bones) from overseas, thanks to his connections in the military. How did he incorporate these objects into his collection? In 1817, Brugmans described his collection to the Leiden governors as part of his offer to sell it. His description reveals that he had classified osteological preparations from foreign countries in a separate category, subdivision 14, which he described as follows:

> Changes in the normal condition and the resulting forms of the animal species. Especially of Man due to climate, way of life, etc. – This includes an extraordinarily rare and important series of approximately 120 human skulls from many different regions, all of them arranged according to their geographical locations, starting with the North Pole and ending with the Equator – Casts of faces of various nations are added to this, etc.[38]

Brugmans was interested in 'changes in the normal condition', because these could help one understand the workings of nature.[39] With regard to the taxonomy of humankind, Brugmans followed German scholar Johann Friederich Blumenbach in dividing the human race into five sub-races.[40] Due to external influences – 'climate, way of life, etc.' – variations and hybrids occurred. Studying these variations would lead to a better understanding of how nature worked in 'normal' cases. Hence, to gain insight into the formation of the five sub-races, it was helpful to study skulls from different nations (and thus, influenced by different external factors). In Brugmans's day, studying skulls usually meant describing individual skulls and using these descriptions to uncover similarities and differences between 'races'.[41]

The physical anthropologists of the second half of the nineteenth century rejected the descriptive approach of Brugmans's time. Instead, they aimed to create a 'scientific' discipline. They strived for conclusions based on a large number of precise numerical measurements; just like researchers in other fields at the time, anthropologists pursued the new scientific ideal of objectivity and they trusted numbers measured mechanically to lead them there.[42] The anthropologists built on Adolphe Quetelet's idea of *l'homme moyen*, the 'average man'. Quetelet, a Belgian astronomer, pioneered the use of statistical methods in the social sciences in the 1830s and 1840s.[43] He focused not on the individual or the particular, but on the whole and the average; an approach that was followed by researchers in many fields, including anthropology. This new 'scientific' approach forced the Leiden armchair anthropologists to get up, take their measuring rods, and reinvestigate the old Brugmans preparations.[44] Brugmans's labels and Sandifort's descriptions no longer sufficed.

Jan van der Hoeven was among the first Leiden researchers to investigate Brugmans's preparations quantitatively. In 1842 he published his book *Bijdragen tot de natuurlijke geschiedenis van den negerstam* (Contributions to the natural history of the negro race). The natural history of humankind, Van der Hoeven explained, belonged to the larger science of anthropology. Its two main research areas were the differences between humankind and the other animals, and the differences between humans, in particular between the different human races. Van der Hoeven focused on the latter, and considered comparing the skulls of different races particularly useful for studying these differences.[45] His book therefore contained a comparison of 'Negro' and 'European' skulls. The comparison was quantitative and based on averages, not individual cases: measurements and statistics formed the foundations of nineteenth-century physical anthropology. The average dimensions of the 'Negro skulls' came from a detailed investigation of ten skulls from the Anatomical Cabinet, all of them from the Brugmans collection. Van der Hoeven admitted that ten was a small number and perhaps insufficient to draw general conclusions.[46] He explained why he had published his findings anyway: he hoped his first results would stimulate other people to collect measurements as well.

Van der Hoeven had carefully measured all ten Brugmans skulls, even though he had already published some of their measurements before.[47] His new measurements had yielded more accurate numbers, something he considered important. He presented his results in a table.[48] For each skull, he provided 12 different dimensions, including the height and the length of the skull, the width of the occipital hole, and the largest distance between the zygomatic arcs. He subsequently averaged these dimensions and compared the results with the average dimensions of European and Chinese skulls. He defended his method as follows:

> If some people think that this average measure is an imaginary thing, we partly agree with them. But it is imaginary in the way that the average temperature, the average barometric pressure, etc., are imaginary. And meanwhile, physicists will not give up these imaginary phenomena; they have learned too many fine and useful things from them. I hope that in the natural history of Man, we will follow our scientific friends in this regard. For more on such research methods, I refer to the shrewd writings of Quetelet.[49]

Van der Hoeven stressed the value of averaging and invoked Quetelet to strengthen his claim – in other words, he was a typical 'scientific' anthropologist.

Van der Hoeven was a professor in natural history at the faculty of natural sciences. The Anatomical Cabinet was part of the medical faculty and was managed by the professor of anatomy. How did Van der Hoeven gain access to the Brugmans skulls? In his book on the 'Negro race', he wrote:

> The Negro skulls that I have investigated for this piece all belong to the collection of Professor Brugmans, which is now in Leiden University's museum of anatomy. The highly-learned Mr Sandifort opened this collection for my research with a willingness for which I wish to thank him publicly.[50]

As a curator of the Anatomical Cabinet, Sandifort had to follow the regulations outlined in the 1815 Royal Decree on Higher Education. The Decree prescribed in detail which professor should manage what collection; the anatomy professor was appointed to manage the Anatomical Cabinet.[51] Other professors could borrow objects for teaching and research purposes with the permission of the managing professor.[52] The 1815 regulations regarding collections were replaced in 1879, when a new decree governing the management and use of 'collections, institutions and teaching aids' in higher education was issued.[53] Again, borrowing objects from the collections was explicitly permitted, as was removing them from the buildings in which they were kept, with the prior consent of the responsible official.[54]

Researchers had access not only to institutional collections, but also to private collections, something that is illustrated by the dissertation that Teunis Zaaijer wrote in 1862.[55] Zaaijer examined two female East-Indian pelvises from the collection of the academic hospital, and compared them with five other pelvises. Four of these belonged to the collection of Amsterdam anatomist Willem Vrolik, who had sent them from Amsterdam to Leiden at the request of Zaaijer's supervising professor, Abraham Simon Thomas.[56] Apparently, collectors were willing to send preparations to other cities to facilitate research.

The fifth comparative preparation Zaaijer used came from the Anatomical Cabinet and belonged to the Brugmans collection. Sandifort described it in the *Museum Anatomicum*, the Cabinet's catalogue, as the 'pelvis of an adult Javanese woman, the bones artificially connected'.[57] He probably based his description on a label or an inscription written by Brugmans, for it is unlikely that he would have made the link to Java had he encountered the pelvis without any description. According to the present-day database of the Anatomical Museum, the pelvis bears the following inscription: 'pelvis feminae adultae javanensis' (pelvis of an adult Javanese woman).[58] This might very well be Brugmans's own inscription. Sandifort's modified description of the Javanese pelvis was further extended by Zaaijer, who explained that the bones were held together with metal wire (copper, according to today's database). More importantly, just like Van der Hoeven had done for the skulls, Zaaijer introduced a quantitative description of the pelvis: he measured 20 dimensions, including the depth of the pelvis at its sides, the width of the pubic arc, and the length of the sacrum. He did the same for the other pelvises he examined and, again like Van der Hoeven, he compiled his results into a table to facilitate comparisons.[59]

The reinterpretation of Brugmans's anthropological preparations did not end with Van der Hoeven and Zaaijer. Although new colonial skulls arrived in large numbers from the 1860s onwards, researchers continued to use skulls from the Brugmans collection. One example can be found in the 1877 dissertation written by Pieter Koning, one of Zaaijer's students. Koning examined Chinese skulls, and although the majority of the 67 skulls he measured had been acquired in recent years, he also used older skulls, including 2 from the Brugmans collection.[60]

All the anatomist-anthropologists working on the Brugmans skulls extracted new information from them. In Rheinberger's terms: they moved back from the epistemologicum, the stabilized object, to the original object of inquiry. This was easy, since the skulls were made of what they represented. With other epistemologica, epistemic recall would have been more troublesome: it would not have been possible with, for instance, drawings of the skulls, such as those published in the fourth volume of the *Museum Anatomicum*. However, many of the measurements could have been made using plaster casts. Although these are not made of what they represent, they do contain all of the necessary information (in this particular instance and for this particular quantitative research question). In the next section, on pathological anatomy, we will encounter different types of reinterpretations, which neither drawings, nor three-dimensional models, would have allowed.

Pathological anatomy

In 1855, Leiden professor Hidde Halbertsma published a treatise on the pathological anatomy of teeth.[61] In his research, he used at least ten dental preparations from the Brugmans collection. He described them microscopically – something that he could do only after having partly dissected the preparations, as he explicitly acknowledged:

> In a few very limited places, the structure of these globes [the *globuli dentis*, thought to be involved in the production of dentine, a component of teeth] is different from that in by far the biggest part of the cross section, from which I have ground microscopic slides.[62]

Halbertsma depicted and described what he saw through his microscope – and in doing so, he reinterpreted a macroscopic pathological preparation on a microscopic level. This practice was not unusual in the mid-nineteenth century; it followed from a shift in pathological theories.

Until the 1750s, medics largely understood disease in terms of Hippocratic interpretations of the movement of fluids through the body. Theories abounded, but all of them highlighted the build-up and balance of the fluids (humours). Moreover, disease was understood holistically, affecting the body as a whole. This changed when a new conception of disease arose: disease as a *localized* entity, caused by changes in a specific body part. The Italian anatomist Giovanni Battista Morgagni advocated this new view early on. In 1761, two years before Brugmans was born, Morgagni published his magnum opus: *De sedibus et causis morborum per anatomen indagatis* (*The Seats and Causes of Diseases, Investigated by Anatomy*). Soon after, medical researchers widely adopted the localized view of disease. Medical doctors treating patients did the same, although they continued to use humoral theories simultaneously.

Brugmans explained diseases locally. Take for example his ideas on cancer, summarized by Abraham Capadose in his eulogy (1825):

> [Brugmans's] explanation of the origin of cancers also belongs to the propositions with which Brugmans tried so vigorously to refute the principles of so-called humoralists; he understood them [cancers] not as being already present in the blood before the vessel system was affected (as was still claimed by the learned Van Gesscher and many distinguished medical men), but as preceded by a peculiar change in the vessels and other solid parts.[63]

According to Brugmans, the cause of cancer was not to be found in one of the humours (here: blood), but in a specific body part (here: the vessels). Note that humoralists did not deny that vessels (or other solid body parts, depending on the type of cancer) were affected in the case of cancer, but they thought the cancer was present in the blood first and then moved to the vessels – the damaged vessels were a *consequence* of the disease, not its *cause*. In their view, the cause was to be found in the humours; and the diseased body part, as a mere consequence, was not their first concern.[64] In the eyes of Brugmans and other followers of Morgagni, to understand disease, one had to study its loci: the diseased body parts. These body parts could be found in pathological collections – a new phenomenon. Until then, early modern anatomists had primarily collected preparations showing the normal (or even the perfect) body. Malformed and diseased body parts were collected from time to time, but mainly as a contrast to the healthy body, not because they were considered interesting in themselves.[65] With the arrival of the localized view of disease, researchers required preparations of pathological body parts.

Which parts of the pathological body one collected, and how one studied them, depended on where one thought disease was localized. For Morgagni in the mid-eighteenth century, disease was primarily localized in organs. In the nineteenth century, however, the loci of disease would become even smaller.[66] Early in the century, following the work of the Frenchman Xavier Bichat, pathologists shifted their focus from organs to tissues. Soon after, in the 1830s and 1840s, the microscope became popular in medicine, leading to a cellular approach to pathology in the second half of the nineteenth century. Researchers now localized disease in cells, and they described it microscopically.

The changing loci did not end the need for pathological collections.[67] After all, diseases were still linked to specific body parts. Furthermore, we should understand the shifts to smaller loci not so much as replacements, but as additions; the larger seats continued to be important, but they were supplemented by descriptions on smaller levels.[68] Hence, older, macroscopic preparations still had their uses, but they had to be supplemented by microscopic preparations and descriptions of the same diseases. Researchers often used existing preparations for microscopic research, because it took a lot of time and effort to build a pathological collection from scratch (there are many diseases, and most bodies – scarce already – tend to display only one of them).

The microscopic reinterpretation of macroscopic preparations was not limited to Leiden, of course. The nineteenth-century pathological catalogues of the London Royal College of Surgeons, for example, mention the microscopic re-examination of older preparations,[69] as do the annual reports of the College's museum curator, such as that from 1890–1891: 'Advantage has been taken of the opportunity presented by the re-mounting of many old preparations to make microscopic sections of all growths not previously examined.'[70] We do not know what 'old' means here, but an early twentieth-century case at the College shows that even preparations that had been made 150 years earlier were reinterpreted. In 1909, curator Arthur Keith received, as he put it, 'permission to cut Hunterian free martin specimens'.[71] A freemartin is the hermaphroditic female calf of a mixed cow-twin. John Hunter had studied the freemartin in the late eighteenth century, and in the early twentieth, Keith wanted to revisit Hunter's freemartin preparations.[72] Hunter had based his ideas mainly on an external investigation of the preparation, but Keith wanted to investigate them microscopically and describe them on a cellular level.[73] Afterwards, he reported back to the Hunterian Trustees (who had granted him permission):

> The specimens you have given me the privilege of examining have been preserved – some of them at least – for over 140 years. It is not necessary to allude to the advantage of being able to verify and augment observations made after so long an interval. The state of preservation of the specimens is so good that there is every reason to believe that some future investigator, in the light of further progress in our knowledge, may still be able to glean fresh information from a re-examination of these specimens.[74]

Keith's investigation was by no means the first time that Hunterian preparations were re-examined; preparations were reinterpreted throughout the nineteenth-century in order to update catalogue descriptions, a process that often necessitated redissection.[75] Thus, old preparations continued to be used even at the College, which in the nineteenth century was known for its flourishing collection, to which thousands of new preparations were added in the course of the century.[76]

Nineteenth-century researchers were not afraid to cut into old preparations, even if these had been made by famous anatomists like John Hunter.[77] Or by Sebald Justinus Brugmans, for that matter, as is evident from Halbertsma's research on teeth. And Halbertsma was not alone in dissecting Brugmans's preparations. Jan Nicolaas Bogtstra and Johannes Boogaard did so as well, about a decade after Halbertsma's work on teeth. They researched a malformation of the skull, for which they used several skulls from the Anatomical Cabinet. Boogaard, Bogtstra's supervisor, wrote an article about this, in which he stated: 'Dr. Bogtstra described five skulls from the Leiden University Anatomical Cabinet [in his dissertation]. All these skulls were sawn through vertically, close to the median plane, in order to simplify the investigation.'[78] Of the five skulls referred to here, two were from the Brugmans collection.[79] At least one skull was sawn through not once, but twice. This follows from Figure 2.2, an illustration from Bogtstra's dissertation, later reprinted in Boogaard's article.

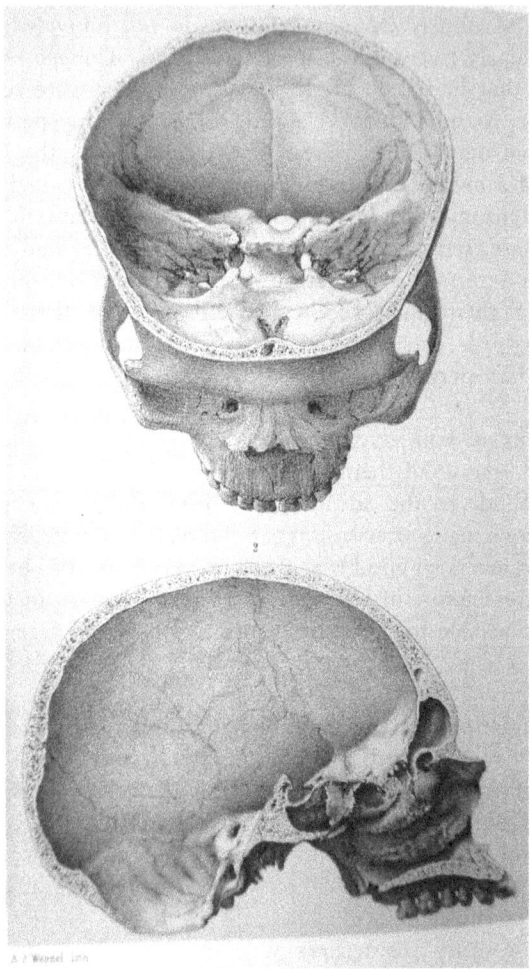

Figure 2.2 Skull collected by Sebald Justinus Brugmans, reinvestigated by doctoral
student Jan Bogtstra in the 1860s. The illustration shows that Bogtstra
had sawn through the skull. 'Horizontale en sagittale coupe van schedel
V', drawing by Mr Hoffmeister in Jan Nicolaas Bogtstra, *De schedel met
de ingedrukte basis* (Leiden: Hazenberg, 1864). Courtesy of Leiden University Library, 240 G 2: 3 qu.

The two images depict the same skull, according to the caption.[80] It was
the skull of a Spanish man, collected by Brugmans. The illustrations were
made by the Leiden illustrator Hoffmeister, who drew them according to
photographs made of the skulls. The drawings show that Bogtstra and
Boogaard must have sawn through the skull both vertically and horizon-
tally – otherwise Hoffmeister would not have been able to draw both sections. Note that he still had to combine two photographs for at least one of

the illustrations. If the skull had been sawn through vertically first, the photograph that was subsequently taken (of one of the resulting halves) could serve as the basis for the bottom illustration. To then create the top figure, it would have been necessary to saw *both* the left and the right half into two, take two photographs of the resulting lower quarters, and then combine them into one horizontal cross section.

The re-examinations described in this section could only have been conducted using preparations, not models or any other epistemologicum. Other epistemologica lack the information (usually: the microscopic structure) that is needed for reinterpretation, because they are not made of what they represent. Preparations are, which makes them remarkably flexible. This flexibility meant that they continued to be useful in medical research for a long time, but it also had a downside: it limited the availability of preparations for *other* users – including researchers outside the medical faculty, as we will see in the next section.

Comparative anatomy

Even after Halbertsma's reorganization in the 1860s, the Anatomical Cabinet still contained animal preparations from the Brugmans collection. And like the human preparations, they continued to be used in research and teaching in a hands-on manner. Halbertsma himself provided an example in his study of hermaphroditism in fish.[81] He was aiming to show that hermaphroditism occurred more often in fish than had so far been assumed. To do so, he described cases of 'abnormal hermaphroditism', rare instances of hermaphroditism in species of fish for which non-hermaphroditism was the norm – as opposed to the then well-known 'normal' hermaphroditism of species in which each fish had both male and female reproductive organs. For one of the cases, Halbertsma used a Brugmans preparation to prove his point.[82] In 1807, when cleaning a bass, a Leiden fishmonger had noticed that the fish had both soft and hard roe. He was, Halbertsma wrote, 'sensible enough' to bring the fish to Brugmans without disturbing it further. Brugmans turned it into a preparation, which Halbertsma now reinvestigated.[83] Halbertsma included an illustration of the bass (see Figure 2.3), which he described as the type of bass endemic to the Netherlands.

In the drawing, four of the fish's internal organs can be seen: the soft roe, the hard roe, the intestinal canal, and the straight intestine. Halbertsma explained that the liver was not visible in the illustration, because it was located behind the bowels and the abdominal wall.[84] But when he argued that this preparation was an example of abnormal hermaphroditism, he also referred to the bass's liver, commenting that it could be readily identified.[85] This suggests that he could see it. But if the liver was hidden from normal view, as (not) shown in the illustration, Halbertsma must have opened the jar and taken out the fish to

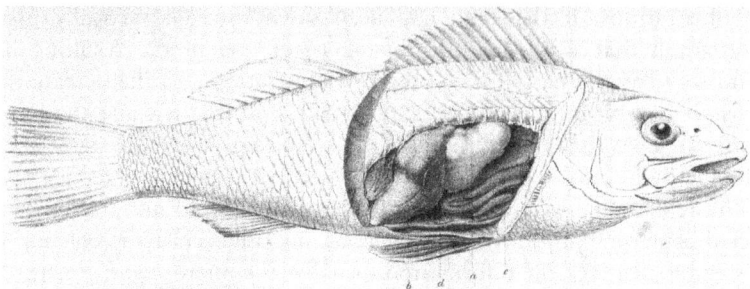

Figure 2.3 Wet preparation of a bass collected by Sebald Justinus Brugmans, reused
by Hidde Halbertsma in his research on hermaphroditism in fish in the
1860s. Lithograph by A. J. Wendel in Hidde Halbertsma, 'Normaal en
abnormaal hermaphroditismus bij de visschen', *Verslagen en mededeelingen der Koninklijke Akademie van Wetenschappen: Afdeeling natuurkunde* 16 (1864): 165–78. Courtesy of Leiden University Library,
Collection Maatschappij der Nederlandse Letterkunde, G 100 3969.

pull away either the bowels or the abdominal wall so that he could see
the liver. Furthermore, he provided precise measurements of the organs,
which would have been difficult had the preparation remained safely in
the jar.[86]

Although the use of animal preparations to study and teach human medicine might seem surprising to us, it was common in the nineteenth century.
All Dutch universities had comparative anatomical preparations in their
medical collections, as required by the Royal Decree on Higher Education.[87]
Medical institutions in other European countries owned animal preparations
as well: they included the medical faculty at the University of Berlin, the
Royal College of Surgeons in London, and the medical faculty at the University of Vienna, where the proper location of the collections was a matter of
fierce debate.[88] The museum of the Karolinska Institute, a medical school in
Stockholm, had a gallery with animal skeletons, including an African ele-
phant, a giraffe, and a walrus.[89] In nineteenth-century Europe, all significant
medical collections kept animal objects alongside human preparations.

Studying the anatomical structures of animals was a regular method for
answering questions about human anatomy and physiology in the nineteenth
century.[90] This method was not new; it had been common in the early
modern period as well. In fact, it was the main reason that Brugmans gave
for building his comparative anatomy collection. In 1807, he explained the
ideas behind his collection in a memorandum:

> The undersigned [Brugmans] [has] devoted himself to building, at a
> high cost and with much work, a rather extensive collection for the
> benefit of his classes in natural history, in particular the ones on com-
> parative anatomy; in order to be able to substitute to a certain extent, in

two or three branches of science, for what is missing in the academic collection, so as not to keep his audience ignorant of the advances that have been made in natural history, in particular as a subsidiary science to the anatomy and physiology of man, by so many famous men all over Europe in the last few years.[91]

Brugmans stated that he had built his collection in particular for his classes in comparative anatomy. In these classes, he wanted to focus on the aspects of natural history (of which he considered comparative anatomy to be a part) that assisted human anatomy and physiology. In other words, his collection was primarily intended not to advance the understanding of animals, but to teach medical students about the human body. I would therefore refer to Brugmans's comparative anatomy as a 'medical' comparative anatomy.

Until the late eighteenth century, this was the main kind of comparative anatomy.[92] But around 1800, a new approach emerged: 'zoological' comparative anatomy.[93] Like medical comparative anatomists, zoological comparative anatomists such as Georges Cuvier (1769–1832) studied animal structures. Their aims differed, however: zoological comparative anatomists did not try to answer research questions on human anatomy and physiology. Instead, they focused on zoological questions; in particular, they compared the anatomical structures of different animals, to discover the natural laws that described these structures.

Professor of natural history Jan van der Hoeven (whose work on 'Negro skulls' has already been discussed) admired Cuvier and practised the new, zoological comparative anatomy. At least, he tried to – but several obstacles prevented him from practising this new discipline as he wanted. A year before his death, he wrote:

> As early as 1829, I pressed the university governors to establish a cabinet of comparative anatomy; I continued to do so until 1861, when I was treated in a way that made me cease my efforts once and for all. I set great store by a collection of comparative anatomy, but even greater store by my independence, and I'd rather abandon my favourite idea than desperately beg for something that science can claim legitimately.[94]

The quotation reveals the main obstacle that Van der Hoeven encountered: he lacked a comparative-anatomical collection. (It also reveals that the relationship between Van der Hoeven and the university governors was tense, to say the least – they continuously refused his requests and finally, in 1861, relocated part of the Brugmans preparations without consulting him.) Like pathological anatomy and physical anthropology, comparative anatomy (whether zoological or medical) was a collection-based research field.[95] If Van der Hoeven wanted to practise comparative anatomy, he needed access to a proper collection. The only comparative anatomy collection present in Leiden at the time (Van der Hoeven was a professor from 1826 to 1868) was the Brugmans collection, which was still largely housed in the Anatomical Cabinet. The anthropological collection was also located in the Cabinet and, as we have seen, Van der Hoeven could easily use preparations from this collection. Thus, what was the problem with the comparative anatomy collection? Could Van der Hoeven not have done the same – simply borrow the preparations he needed? The answer is no, he could not. Van der Hoeven was trying to achieve something different: instead of answering research questions (as with his anthropological work), he was aiming to establish an independent research field – a new discipline, one might say.

Van der Hoeven expressed his ambition to establish comparative anatomy as an independent area of study in his 1867 article 'Over den aard en het doel der vergelijkende ontleedkunde, en over hare hulpmiddelen te Leiden' (On the nature and the purpose of comparative anatomy, and on its resources in Leiden).[96] In the opening paragraph, he announced his intention to boost the reputation of comparative anatomy in the Netherlands, which, so he stated, was sadly wanting.[97] He also repeatedly stressed that comparative anatomy needed to be independent of medicine. His ambition also followed from his attempts to recruit practitioners. No discipline can do without practitioners, and Van der Hoeven hoped to create them by tempting medical students into comparative anatomy, as he revealed in his letter to Collot d'Escury: 'It is bound to work; the zeal of many for this field of study [natural history based on comparative anatomy] will be aroused and the science will attract more and more practitioners.'[98]

Collections can play a key role in discipline formation. Frances Larson has shown how the acquisition of the Pitt Rivers collection laid the foundations for the discipline of anthropology at Oxford University.[99] The collection visibly demarcated the boundaries of the young discipline.[100] Furthermore, the presence of the collection forced the university's administrators to allocate funds to anthropology: they could not allow the prestigious collection to deteriorate. An independent comparative anatomy collection, managed by Jan van der Hoeven, could have improved the position of comparative anatomy in Leiden in a similar way. It would have provided Van der Hoeven with some financial backing from the administrators, as well as a collection space, which could have been used for research. Furthermore, the collection in itself, with its thousands of preparations, would have given the discipline a strong, visible presence. Once big enough, a collection can gain a kind of momentum of its own, which ensures its continued existence and further growth; for example, because individuals start donating their objects to the collection. Cuvier admitted that he used his comparative-anatomical collection in this way: 'I succeeded in making my collection so important that soon nobody dared to oppose its further enlargement.'[101]

Van der Hoeven wanted a collection to establish comparative anatomy as an independent field of study. Although the Brugmans collection contained over 2,000 comparative anatomy preparations, it could not fulfil this role. There were two problems. The first, which related to the collection's contents, could have been overcome, had it not been for the second: the prolonged use of the collection in the medical faculty.

Let us consider the contents first. Brugmans built his collection to support his medical teaching. How did this intention shape his collection? Which animal structures were needed to learn about the human body? According to Brugmans (and almost everyone else), the structures of the animals closest to humans: the vertebrates. Within the vertebrates, mammals were considered the most useful. Indeed, vertebrates in general, and mammals in particular, were best represented in Brugmans's collection. Almost all of the comparative anatomy preparations involved vertebrates, with only 71 invertebrates.[102] Moreover, more than half of all objects stemmed from mammals (1,198), of which almost a third were human preparations. This made the collection useful in medical teaching, as was happily acknowledged by medical professor Gerard Sandifort:

> If one takes a look at the sketch outline of this collection, drawn up by the late professor Brugmans himself ... it will soon become clear ... that from the very beginning the intention behind the collection was to gain more knowledge about the structure and the actions of the human body, and all of the late Professor Brugmans's classes on comparative anatomy also had this intention ... In the present-day state of this science [physiology] it is not possible to explain the various functions of the parts of the human body without resorting to comparative anatomy, this being a rich resource for physiological knowledge of the human body.[103]

Van der Hoeven, on the other hand, was not all that happy about the composition of the Brugmans collection. In his eyes, a comparative anatomy collection aimed at medical teaching could never suffice when teaching the new and 'real' comparative anatomy. He argued that a professor in zoology and comparative anatomy could not be expected to make do with the medical faculty's comparative anatomy collection, just as a chemistry professor could not be expected to borrow the preparations he needed from the professor in *materia medica*.[104] In his letter to governor Collot d'Escury, Van der Hoeven referred to the difference between the two types of collections:

> A collection of comparative anatomy as an appendage to a cabinet of human anatomy and physiology, no matter how excellent, never *could*, nor *should*, be arranged like a collection of comparative anatomy used for the explanation of zoology. The latter, however much it is instrumental in general physiology because of the mutual ties that connect all sciences, has to have an extensiveness that also has a completely independent tenor.[105] (italics in the original)

In other words, medical comparative anatomy collections required only preparations that would help answer questions about human anatomy and physiology; a zoological comparative anatomy collection needed much more. According to Van der Hoeven, the aim of zoological comparative anatomy was to formulate 'a theory of animal forms', that is, an explanation of why animals (including man) are built the way they are. This explanation could be achieved by comparing the structures of different animals, for this would result in a classification of 'all typical varieties'.[106] This is most easily done when the collection is arranged according to species, whereas the study of animal anatomy to understand human medicine is best served by an arrangement according to organ systems (lungs with lungs, muscles with muscles, and so on). Furthermore, to formulate a theory of animal forms, comparative anatomists had to study *all* types of animals, not just the vertebrates. This was the 'extensiveness' Van der Hoeven referred to in his letter; an extensiveness that the Brugmans collection lacked due to its limited number of invertebrates.

Although the Brugmans collection itself was unsuitable for researching and teaching the new zoological comparative anatomy, it could have formed a foundation for a collection that was suitable. At the Royal College of Surgeons in London, comparative anatomist Richard Owen had transformed John Hunter's collection of animal preparations from a collection intended for the study of human medicine into a collection for the study of Cuvierian comparative anatomy (rebranding John Hunter as the father of the new comparative anatomy in the process).[107] Van der Hoeven, like many of his contemporaries, admired the College's Museum, 'where everything has been arranged in the finest order, ascending from the least complex and least organized [*bewerktuigd*] creatures to mankind'.[108] Although Van der Hoeven had to work on a much smaller scale, this was the type of collection that he was aiming for. He thought the Brugmans collection could serve as a starting point, as he made clear in his continuous requests for a separate comparative anatomy collection.[109] In 1859, for example, he proposed to merge the Brugmans collection (that is, the comparative anatomy part) with his private collection. Together, the collections would form the starting point for an institutional comparative anatomy collection, which, in time, could be extended further. The governors refused all of his requests and Brugmans's preparations remained in the Anatomical Cabinet; that is, until 1861, when the governors transferred part of the comparative anatomy preparations from the Cabinet to the Museum for Natural History – all behind Van der Hoeven's back.[110] Once in the Museum, the preparations were even less accessible to Van der Hoeven than in the Cabinet, because of his fierce conflict with the museum director, Hermann Schlegel.[111] The governors' move proved too much for Van der Hoeven, and he gave up his 30-year quest for an independent comparative anatomy collection. During that quest, Van der Hoeven had never understood – or so he claimed – why the Brugmans preparations had been placed in the Cabinet to begin with, a decision he called 'inexplicable'.[112] But from our perspective, it seems quite simple. Eighteenth-century comparative anatomy preparations belonged in a nineteenth-century medical collection because animal preparations were widely used in nineteenth-century medicine; and old preparations, being made of what they represented, could easily be adapted to new research questions.

Van der Hoeven never established the independent comparative anatomy collection he desired. The composition of the Brugmans collection was wrong, but that problem could have been overcome, had it not been for the second problem: the collection's location in the Anatomical Cabinet. As long as the collection was housed in the Cabinet, Van der Hoeven could never exert the necessary influence on the collection to alter its contents. Nor could the collection play its required role as the visible manifestation of the new, independent research area of zoological comparative anatomy. Before Van der Hoeven could use the Brugmans collection to demarcate the discipline of comparative anatomy, he had to gain control over the collection. He tried to do so, more than once, but he failed. In the end, some of Brugmans's comparative anatomy preparations ended up in the Museum for Natural History; many remained in the medical faculty's Anatomical Cabinet throughout the nineteenth century. The Cabinet's curators, all medical professors, were unwilling to dispose of them, because they continued to be useful in medical research and teaching. Since they enabled reinterpretation quite well, the preparations could be adapted to changing practices and theories; not only in comparative anatomy, but also, as we have seen, in other areas of study.

Conclusion

For nineteenth-century researchers, preparations were flexible objects. They were made of what they represented and thus enabled, in Rheinberger's terms, epistemic recall. This helps explain their prolonged use. But we should not forget that although they are made of what they represent, preparations are *made* nonetheless – preparations are *not* naturalia. And thus, their reinterpretation is not limitless.

Rheinberger points to the limits of reinterpretation when he claims that epistemic recall is easier with herbarium plants than macroscopic preparations, because herbarium plants have been manipulated less.[113] Macroscopic preparations undoubtedly contain a fair amount of manipulation: a kidney preparation is not *solely* made of kidney, but also contains materials such as injection mass and preparation fluid, and a great deal of work. And indeed, this may complicate its reinterpretation. Part of the information present in the raw material inevitably gets lost in the making – or the keeping; something that Hidde Halbertsma discovered when working on the hermaphrodite bass. Halbertsma was unable to complete his reinvestigation because the preparation fluid had affected the fish's organs:

In our preparation, it could, to our regret, no longer be demonstrated how the seed was ejected, because the deeper-lying organs were in a softened condition and hence the probable *vas deferens* [the duct carrying the spermatozoa from the epididymis] could no longer be detected.[114]

Present-day biologists also encounter such problems when attempting to extract DNA from old preparations stored in formaldehyde, which is less DNA-friendly than alcohol.

The nineteenth-century afterlife of the Brugmans collection has shown that reinterpreting macroscopic preparations is possible, but the process has its limits. Nonetheless, the possibilities were large enough to keep nineteenth-century medical researchers using eighteenth-century preparations throughout the century. As a result, researchers from outside the medical faculty could be excluded, as was the case with Jan van der Hoeven. Other user groups were excluded as well, groups not involved in research at all: lay visitors and the university governors. It is to them we now turn.

Notes

1   Sandifort, *Museum Anatomicum 3*, xxiv.
2   Theunissen, *Nut*, 42.
3   Elshout, *Leidse kabinet*, 107.
4   These numbers are rough estimates, because no complete catalogues or inventories were kept. I have based the number of 8,000 on the four volumes of the major catalogue *Museum Anatomicum* (7,382 preparations listed, all created before 1815) and an estimate of the number of preparations acquired in the first half of the nineteenth century. In addition to the Brugmans and the Bonn collections (both catalogued in the Museum), three major collections were acquired: Jacobus Rocquette's (doctor and lecturer in Haarlem; collection acquired in 1818); Ledeboer's (first name and occupation unknown; collection acquired before 1827); and Simon du Pui's (Leiden professor; collection acquired in 1837). Du Pui's collection contained 76 preparations (Elshout, 24–25). The sizes of the other collections are unknown. Gerard Sandifort considers them less important than the collections of Brugmans and Bonn (Sandifort, *Museum Anatomicum 3*, Praefatio, 3–4), which suggests they were smaller. Therefore, I have estimated that they contained a few hundred preparations. The annual reports regularly mention individual additions to the collections; in my estimate, 250 preparations in total. Note that my numbers might be too high, because some of the preparations are mentioned twice in the *Museum*'s volumes (though other descriptions probably included more than one preparation) and because part of the preparations described in the *Museum*'s first two volumes were destroyed by an exploding gunpowder ship in 1807. Even if the actual numbers were lower, however, my claim that most of the preparations were made before 1815 still holds true.
5   See for example Alberti, 'Owning and Collecting', on British collections.
6   On comparative anatomy, see Nyhart, *Biology Takes Form*. On pathological anatomy, see Maulitz, *Morbid Appearances*. On embryology, see Hopwood, 'Embryology'.
7   Cunningham, 'Old Physiology'; Cunningham, 'Old Anatomy'.

8 On the rise of the laboratory in medicine, see Cunningham and Williams, *Laboratory Revolution*. On the birth of the clinic, see Ackerknecht, *Paris Hospital*; Foucault, *Birth of the Clinic*. On the growing importance of microscopy, see Schickore, *Microscope*. On the construction of cell theory, see Harris, *Birth of the Cell*.

9 Beukers, 'Beginjaren van de microscopie'; Beukers, 'Groei en ontwikkeling', 198.

10 Rheinberger, *Epistemology of the Concrete*, 233–34.

11 Rheinberger, 'Präparate', 9–10. Rheinberger uses philosopher of science Bas van Fraassen's definition of representation.

12 Rheinberger, *Epistemology of the Concrete*, 238.

13 Biographical information on Brugmans can be found in the many obituaries that appeared after his death, partly listed in Wallé, *Leiden Medical Professors*, 130–31. The most extensive are those by H. C. van der Boon Mesch and Abraham Capadose, written for a prize essay competition organized by the Hollandsche Maatschappij der Wetenschappen (Holland Society of Sciences and Humanities). Van der Boon Mesch, 'Lofrede'; Capadose, 'Lofrede'. For a more recent description of Brugmans's life, see De Jonge, 'Sebald Justinus Brugmans', 1999, and De Jonge, 'Sebald Justinus Brugmans', 2001. Brugmans's correspondence has been described in Van Heiningen, *Correspondence of Sebald Justinus Brugmans*.

14 Brugmans, *Dissertatio*.

15 Sandifort, *Museum Anatomicum 3*, xiii.

16 De Jonge, 'Sebald Justinus Brugmans', 1999, 10.

17 The collection is catalogued in Sandifort, *Museum Anatomicum 3*. For a list of visitor reports and other literature on Brugmans's collection, see Engel, *Hendrik Engel's Alphabetical List*, 46. Also helpful is the description by Cornelis van der Klaauw: Van der Klaauw, 'Verzameling'.

18 Brugmans et al., *Pharmacopoea Batava*.

19 Van der Klaauw, 'Verzameling', 50–51.

20 De Jonge, 'Sebald Justinus Brugmans', 1999, 46. Sandifort, *Museum Anatomicum 3* sometimes mentions donors in the descriptions, but he tends to keep silent about the objects' provenance. Marieke Hendriksen offers a detailed discussion of the possible provenance of a set of colonial fetuses in Brugmans's collection; see Hendriksen, *Elegant Anatomy*, 144–77.

21 Brugmans to governors, 4 April 1817, file 70, document 56, Archief van Curatoren 1815–1877 (hereafter cited as AC2), Leiden University Library.

22 'Organiek Besluit Hooger Onderwijs', 2 August 1815 (hereafter cited as RDHE 1815), article 177.

23 Of course, the numbers depend on how one classifies the preparations. I have taken the numbers from Van der Klaauw, 'Verzameling', which is based on Sandifort, *Museum Anatomicum 3*. Note that the pathological preparations contain a lot of animal preparations as well, and that the comparative-anatomical preparations include foreign (human) skulls, which could equally be considered a separate category.

24 Minutes of the governors, 22 May 1817, file 3, AC2.

25 C. M. van Dam (widow of Sebald Justinus Brugmans) to governors, 14 October 1819, file 72, document 131, AC2; Minutes of the governors, 25 October 1819, file 5, AC2; Minister of Education to governors, 6 November 1819, file 72, document 141, AC2.

26 Minutes of the governors, 27 November 1819, file 5, AC2.

27 Sandifort to governors, 2 December 1819, file 72, document 149, AC2.

28 Minutes of the governors, 30 September 1820, file 6, AC2; Sandifort to governors, 29 September 1820, file 73, document 124, AC2.

29 Sandifort to governors, 21 October 1820, file 73, document 142, AC2.

30   'Extract, uit het Register der Handelingen en Resolutien van den Minister, voor het Publieke Onderwijs, de Nationale Nijverheid en de Kolonien', No 3, 31 December 1820, cited in Gijzen, *'s Rijks Museum*, 17.
31   Van der Klaauw, *Hooger onderwijs*, 12.
32   De Jonge, 'Sebald Justinus Brugmans', 2001, 6.
33   De Jonge, 'Macht, machinaties en musea', 197.
34   On physical anthropology, see Fenneke Sysling's overview of early Dutch physical anthropology. Sysling, *Racial Science*, 1–46.
35   Sysling, 28–42. Around 1900, physical anthropology's momentum would move again, this time to Amsterdam – with the university's anatomical collections and the newly founded Colonial Institute (1910) as its loci.
36   Sysling, 'Archipelago of Difference', 23.
37   Teunis Zaaijer, 'Katalogus der ras-schedels, bekkens en skeletten in het Anatomisch Kabinet der Rijks-Universiteit te Leiden', 1893, p. 11, archives Anatomisch Museum (no inventory number), Leiden University Medical Center. After 1860, imports from overseas rose quickly. See Sysling, *Racial Science*, 28–34.
38   Brugmans to governors, 4 April 1817, file 70, document 56, AC2.
39   Brugmans to governors, 4 April 1817, file 70, document 56, AC2.
40   De Jonge, 'Sebald Justinus Brugmans', 2001, 22; De Jonge, 'Sebald Justinus Brugmans', 1999, 41–44.
41   Sysling, 'Archipelago of Difference', 14–15.
42   Sysling, *Racial Science*, 3–4. On the rise of statistics in general, see Porter, *Rise of Statistical Thinking*. On the rise of statistics in Dutch medicine, see Klep and Kruithof, 'Quantification and Statistics'. On scientific objectivity, see Daston and Galison, *Objectivity*.
43   Vanpaemel, 'Het getal regeert'.
44   Elshout claims that Brugmans measured his skull preparations, but her source is unclear (Elshout, *Leidse kabinet*, 107). She refers to a catalogue on racial skulls by Sandifort (*Tabulae craniorum diversarum nationum*) which does, indeed, contain some measurements of skulls collected by Brugmans, but Sandifort does not claim Brugmans took these measurements himself. It is more likely that Sandifort measured the skulls, especially given that the measurements are lacking in older catalogues (Sandifort, *Museum Anatomicum 3*; Sandifort, *Museum Anatomicum 4*).
45   Van der Hoeven, *Bijdragen*, 5.
46   Van der Hoeven, 25.
47   Van der Hoeven, section 'Voorberigt' (unnumbered page).
48   Van der Hoeven, 30.
49   Van der Hoeven, 36–37.
50   Van der Hoeven, 26. Other institutional collections, both inside and outside Leiden, were used in research as well. See for example Swaving, 'Eerste bijdrage', 285 (a skull from the Bataafsch Genootschap [Batavian Society]).
51   RDHE 1815, article 178.
52   RDHE 1815, article 200.
53   'Reglement op het beheer en het gebruik der verzamelingen, inrigtingen en hulpmiddelen voor het onderwijs aan de Universiteiten des Rijks', 31 December 1879 (hereafter cited as Regl. 1879).
54   Regl. 1879, articles 6 and 7.
55   Another example can be found in Hidde Halbertsma's article on the third joint on the occipital bone, in which Halbertsma used skulls from the private collections of both Van der Hoeven and Cornelis Swaving. (Halbertsma, 'Derde gewrichtsknobbel', 222.) Swaving himself used skulls from other private collections in his work, see for example Swaving, 'Eerste bijdrage'.

56 Zaaijer, *Twee vrouwenbekkens*, 11.
57 Sandifort, *Museum Anatomicum 3*, 109 (object 1860). The Latin reads: 'Pelvis ossa artificialiter nexa foeminae [*sic*] adultae javanensis.'
58 In the database, the preparation can be found as number Af0168.
59 Zaaijer, *Twee vrouwenbekkens*, table after p. 30.
60 Koning, *Chineesche schedels*, 6–7. On reuse of older preparations by Dutch physical anthropologists in the late nineteenth and early twentieth centuries, see also Sysling, 'Provenance', 204–6.
61 Halbertsma, *Ziektekundige ontleedkunde der tanden*.
62 Halbertsma, 14.
63 Capadose, 'Lofrede', 602. Unfortunately, we have to rely on Capadose's account of Brugmans's ideas in this area, because Brugmans himself did not publish them.
64 Note that many different kinds of humoralists existed; most of them had more complex ideas on disease than Capadose suggests in his eulogy. He tended to oversimplify the view of the humoralists in order to sharpen the contrast between them and Brugmans.
65 On the changing position of pathological preparations in the Leiden collections, see Hendriksen, *Elegant Anatomy*, 108–43.
66 On the development of pathology in the nineteenth century, see Maulitz, *Morbid Appearances*.
67 On pathological collections in the nineteenth century, see Alberti, *Morbid Curiosities*.
68 Maulitz, 'Pathology', 369.
69 See for example Paget, *Descriptive Catalogue*: preface and individual object descriptions (e.g., number 3589, p. 29).
70 Annual report of the conservator to the museum committee 1890–1891, 29 June 1891, file 8/2/2, p.2, Papers of the Hunterian Museum and the Wellcome Museum (hereafter cited as RCS-MUS), Royal College of Surgeons Archives, RCS-MUS, London.
71 Arthur Keith, Work diary 1908–1909, entry 9 February 1909, file 3/1/3, vol. 1, Arthur Keith Papers, Royal College of Surgeons Archives, MS0018, London.
72 Hunter, *Observations*, 46–62.
73 Minutes of the Hunterian Trustees, 10 February 1909, Hunterian Trustees Minute Book Vol. 3, 1900–1959, file 1/1/3, RCS-MUS.
74 Minutes of the Hunterian Trustees, 10 November 1909, Hunterian Trustees Minute Book Vol. 3, 1900–1959, file 1/1/3, RCS-MUS.
75 See for example Cobbold, *Specimens of Entozoa*, iv.
76 On the reuse of pathological preparations with the help of new techniques, see also Claes, 'Nobody's Dead', 200–201, on the Belgian case.
77 Note that Arthur Keith eventually reported to the Trustees that he had managed to re-examine the Hunterian freemartin preparations without damaging them – usually, this was not the case when preparations were reinterpreted microscopically. (Minutes of the Hunterian Trustees, 10 November 1909, Hunterian Trustees Minute Book Vol. 3, 1900–1959, file 1/1/3, RCS-MUS.)
78 Boogaard, 'Indrukking', 86.
79 Bogtstra, *Schedel*, 10–11. After Bogstra had finished his dissertation, Boogaard found three more skulls with the same malformation, which he also re-examined. All three of them belonged to the Brugmans collection (Boogaard, 'Indrukking', 92–93).
80 Bogtstra, *Schedel*, 37.
81 Halbertsma, 'Hermaphroditismus'.
82 Halbertsma, 173–75. The preparation was listed in Sandifort, *Museum Anatomicum 3*, 31; object number 395 of 'Pars prior' of the Brugmansiana. The present-day catalogue number is Af0055.

83   Halbertsma, 'Hermaphroditismus', 173.

84   Halbertsma, 178.

85   Halbertsma, 174.

86   Halbertsma, 173–74.

87   RDHE 1815, article 177.

88   On the collections in Berlin (including that of Rudolf Virchow), see Matyssek, *Rudolf Virchow*, 53; Virchow, *Das neue Pathologische Museum* (reprinted in, Matyssek, *Rudolf Virchow*, 141–58), in particular the floor plan of the university's new pathological museum. On animal preparations in German medical collections, see also Nyhart, *Biology Takes Form*. For the composition of the College's collection, see its many nineteenth-century catalogues (e.g., Royal College of Surgeons, *Descriptive and Illustrated Catalogue*); on its animal preparations, see also Jacyna, 'Images of John Hunter'; Desmond, *The Politics of Evolution* – although both seem to consider them an anomaly, not a common feature of medical collections at the time; a recurring problem in the historiography of nineteenth-century anatomical collections. On animal preparations in British medical collections in general, see Alberti, *Morbid Curiosities*, 57. On comparative anatomy in Vienna, see Buklijas, 'Dissection', 146–51.

89   Åhrén, 'Making Space', 106.

90   In her book on the study of form in nineteenth-century Germany, Lynn Nyhart has shown how morphology – a combination of comparative anatomy, embryology, and histology – was part of both the medical and the zoological faculties at German universities. She argues that morphology's position within German medical faculties changed: the new physicalist physiologists disapproved of morphology, and the field then found a new home in anatomy departments (Nyhart, *Biology Takes Form*). Laurens de Rooy has shown that Dutch anatomists also embraced the study of animal structure in the nineteenth century. He argues that Dutch anatomists used evolutionary morphology, of which comparative anatomy was a part, to escape the crisis that had hit their discipline in the 1860s (De Rooy, *Snijburcht*).

91   Brugmans, 3 March 1807, in a memorandum attached to the report 'Verslag van den Senaat aan Curatoren over het aantal studenten en het gegeven onderwijs' (13 March 1807), cited in Molhuysen, *Bronnen*, 7:90*.

92   On this kind of comparative anatomy in the eighteenth century, and in particular on why it cannot be considered an independent discipline, see Cunningham, *The Anatomist Anatomis'd*, 295–359.

93   On this new kind of comparative anatomy, see Cunningham, 375–79; Cunningham, 'Quis Custodiet Ipsos Custodes?'

94   Van der Hoeven, 'Vergelijkende ontleedkunde', 664.

95   Pickstone, 'Museological Science?', 117.

96   Van der Hoeven, 'Vergelijkende ontleedkunde'.

97   Van der Hoeven, 657.

98   Van der Hoeven to Collot d'Escury, 21 January 1829, file 607, folder E, document 2, AC2.

99   Larson, 'Anthropological Landscaping'.

100 On collections as tools in boundary work, see also Whitehead, *Museums*; Alberti, *Nature and Culture*.

101 Flourens, *Recueil*, 1:183; translation taken from Outram, *Georges Cuvier*, 176. On Cuvier's use of his collection to make comparative anatomy visible, see Outram, 175–80.

102 Van der Klaauw, 'Verzameling', 52.

103 Sandifort to governors, 2 December 1819, file 72, document 149, AC2. In the first sentence, Sandifort refers to the report Brugmans sent to the governors when he offered them his collection.

104 Van der Hoeven to Collot d'Escury, 21 January 1829, file 607, folder E, document 2, AC2.
105 Van der Hoeven to Collot d'Escury, 21 January 1829, file 607, folder E, document 2, AC2.
106 Van der Hoeven, 'Vergelijkende ontleedkunde', 16.
107 Cunningham, 'Quis Custodiet Ipsos Custodes?'
108 Van der Hoeven, 'Vergelijkende ontleedkunde', 662.
109 On these requests, see Van der Klaauw, *Hooger onderwijs*, 10–12; De Jonge, 'Macht, machinaties en musea', 185.
110 Van der Klaauw, *Hooger onderwijs*, 11–12.
111 More on the difficult relationship between Van der Hoeven and Schlegel can be found in De Jonge, 'Macht, machinaties en musea'.
112 Van der Hoeven, 'Vergelijkende ontleedkunde', 663.
113 Rheinberger, *Epistemology of the Concrete*, 238. Herbarium plants are one of the four types of preparations distinguished by Rheinberger. The other three are: anatomical preparations, microscopic preparations, and chromatograms. Chromatograms reveal (and simultaneously are) the components of molecules, the most famous example being DNA barcodes. They are the icons of twentieth-century molecular biology (and twenty-first-century crime shows), but of no concern to us here. To Rheinberger, 'anatomical preparations' are human or animal preparations visible to the naked eye. Note this differs slightly from my usage. For me, 'anatomical preparations' can include microscopic preparations as well; I use 'macroscopic preparations' to define the objects Rheinberger calls 'anatomical preparations'.
114 Halbertsma, 'Hermaphroditismus', 174–75.

Bibliography

Manuscript sources

Leiden University Library, Special Collections: Archief van Curatoren 1815–1877.
Leiden University Medical Center: Archives Anatomisch Museum.
London, Royal College of Surgeons Archives: MS0018, Arthur Keith Papers; RCS-MUS, Papers of the Hunterian Museum and the Wellcome Museum.

Printed sources

Ackerknecht, Erwin H. *Medicine at the Paris Hospital, 1794–1848.* Baltimore: Johns Hopkins University Press, 1967.
Åhrén, Eva. 'Making Space for Specimens: The Museums of the Karolinska Institute, Stockholm'. In *Medical Museums: Past, Present, Future*, edited by Samuel J. M. M. Alberti and Elizabeth Hallam, 102–15. London: Royal College of Surgeons of England, 2013.
Alberti, Samuel J. M. M. 'Owning and Collecting Natural Objects in Nineteenth-Century Britain'. In *From Private to Public: Natural Collections and Museums*, edited by Marco Beretta, 141–54. Sagamore Beach: Science History Publications, 2005.
Alberti, Samuel J. M. M. *Nature and Culture: Objects, Disciplines and the Manchester Museum.* Manchester: Manchester University Press, 2009.
Alberti, Samuel J. M. M. *Morbid Curiosities: Medical Museums in Nineteenth-Century Britain.* Oxford: Oxford University Press, 2011.

Beukers, Harm. 'De beginjaren van de microscopie aan de geneeskundige faculteiten te Utrecht en Leiden'. *Tijdschrift voor de geschiedenis der geneeskunde, natuurwetenschappen, wiskunde en techniek* 6(1983): 65–81.

Beukers, Harm. 'Groei en ontwikkeling: De Leidse faculteit der geneeskunde in het derde kwart der negentiende eeuw'. In *Een universiteit herleeft*, edited by Willem Otterspeer, 76–102. Leiden: Brill, 1984.

Bogtstra, Jan Nicolaas. *De schedel met ingedrukte basis*. Leiden: Hazenberg, 1864.

Boogaard, Johannes A. 'De indrukking der grondvlakte van den schedel door de wervelkolom, hare oorzaken en gevolgen'. *Nederlands tijdschrift voor geneeskunde* 9(1865): 81–108.

Boon Mesch, H. C. van der. 'Lofrede op Sebald Justinus Brugmans'. *Werken der Hollandsche Maatschappij van Fraaije Kunsten en Wetenschappen* 7(1825): 197–360.

Brugmans, Sebaldus Justinus. *Dissertatio physico mineralogica inauguralis de lapidibus et saxis agri Groningani*. Groningen: Doekema & Muller, 1781.

Brugmans, Sebald Justinus, P. Driessen, G. Vrolik, J. R. Deiman, and G. G. ten Haaff. *Pharmacopoea Batava*. Amsterdam: Johannes Allart, 1805.

Buklijas, Tatjana. 'Dissection, Discipline and Urban Transformation: Anatomy at the University of Vienna 1845–1915'. PhD diss., University of Cambridge, 2005.

Capadose, Abraham. 'Lofrede op Sebald Justinus Brugmans'. *Werken der Hollandsche Maatschappij van Fraaije Kunsten en Wetenschappen* 7(1825): 363–605.

Claes, Tinne. 'Nobody's Dead: The Trajectories of the Corpse in Belgian Anatomy, ca. 1860–1914'. PhD diss., KU Leuven, 2017.

Cobbold, Thomas S. *Catalogue of the Specimens of Entozoa in the Museum of the Royal College of Surgeons of England*. London: Hardwicke, 1866.

Cunningham, Andrew. 'The Pen and the Sword: Recovering the Disciplinary Identity of Physiology and Anatomy before 1800. I: Old Physiology – the Pen'. *Studies in History and Philosophy of Biological and Biomedical Sciences* 33 (2002): 631–65. https://doi.org/10.1016/S1369-8486(02)00023-7.

Cunningham, Andrew. 'The Pen and the Sword: Recovering the Disciplinary Identity of Physiology and Anatomy before 1800. II: Old Anatomy – the Sword'. *Studies in History and Philosophy of Biological and Biomedical Sciences* 34(2003): 51–76. https://doi.org/10.1016/S1369-8486(02)00069-9.

Cunningham, Andrew. *The Anatomist Anatomis'd: An Experimental Discipline in Enlightenment Europe*. The History of Medicine in Context. Farnham: Ashgate, 2010.

Cunningham, Andrew. 'Quis Custodiet Ipsos Custodes? Or, What Richard Owen Did to John Hunter's Collection'. In *The Fate of Anatomical Collections*, edited by Rina Knoeff and Robert Zwijnenberg, 23–52. The History of Medicine in Context. Farnham: Ashgate, 2015.

Cunningham, Andrew, and Perry Williams, eds. *The Laboratory Revolution in Medicine*. Cambridge: Cambridge University Press, 1992.

Daston, Lorraine, and Peter Galison. *Objectivity*. New York: Zone Books, 2007.

Desmond, Adrian. *The Politics of Evolution: Morphology, Medicine, and Reform in Radical London*. Chicago: University of Chicago Press, 1989.

Elshout, Antonie M. *Het Leidse kabinet der anatomie uit de achttiende eeuw: De betekenis van een wetenschappelijke collectie als cultuurhistorisch monument*. Leiden: Universitaire Pers Leiden, 1952.

Engel, Hendrik. *Hendrik Engel's Alphabetical List of Dutch Zoological Cabinets and Menageries*. 2nd ed. Prepared by Pieter Smit, with the assistance of A. P. M. Sanders and J. P. F. van der Veer. Amsterdam: Rodopi, 1986.

Flourens, P. *Recueil des éloges historiques lus dans les séances publiques de l'Académie des Sciences*. Vol. 1. Paris: Garnier, 1856.

Foucault, Michel. *The Birth of the Clinic: An Archaeology of Medical Perception*. Translated by A. M. Sheridan. London: Tavistock, 1976.

Gijzen, Agatha. *'s Rijks Museum van Natuurlijke Historie, 1820–1915*. Rotterdam: Brusse, 1938.

Halbertsma, Hidde Justusz. *Bijdrage tot de ziektekundige ontleedkunde der tanden*. Amsterdam: Van der Post, 1855.

Halbertsma, Hidde Justusz. 'Normaal en abnormaal hermaphroditismus bij de visschen'. *Verslagen en mededeelingen der Koninklijke Akademie van Wetenschappen: Afdeeling natuurkunde* 16(1864): 165–78.

Halbertsma, Hidde Justusz. 'De derde gewrichtsknobbel (condylus tertius) van het achterhoofdsbeen'. *Nederlandsch tijdschrift voor geneeskunde* 9(1865): 222–27.

Harris, Henry. *The Birth of the Cell*. New Haven: Yale University Press, 1999.

Heiningen, Teunis Willem van, ed. *The Correspondence of Sebald Justinus Brugmans (1763–1819)*. Tools and Sources for the History of Science in the Netherlands 1. The Hague: Dutch History of Science Web Centre, 2008. www.dwc.knaw.nl/literatuur/tools-sources.

Hendriksen, Marieke M. A. *Elegant Anatomy: The Eighteenth-Century Leiden Anatomical Collections*. History of Science and Medicine Library 47. Leiden: Brill, 2015.

Hoeven, Jan van der. *Bijdragen tot de Natuurlijke geschiedenis van den negerstam*. Leiden: Luchtmans, 1842.

Hoeven, Jan van der. 'Over den aard en het doel der vergelijkende ontleedkunde, en over hare hulpmiddelen te Leiden'. *Nederlandsch tijdschrift voor geneeskunde* 11 (1867): 657–65.

Hopwood, Nick. 'Embryology'. In *The Cambridge History of Science*. Vol. 6, *The Modern Biological and Earth Sciences*, edited by Peter J. Bowler and John V. Pickstone, 285–315. Cambridge: Cambridge University Press, 2009.

Hunter, John. *Observations on Certain Parts of the Animal Oeconomy*. London, 1786.

Jacyna, L. Stephen. 'Images of John Hunter in the Nineteenth Century'. *History of Science* 21(1983): 85–108.

Jonge, Hans de. 'Sebald Justinus Brugmans (1763–1819)'. Master's thesis, Leiden University, 1999.

Jonge, Hans de. 'Sebald Justinus Brugmans: Vergeten geleerde, verlicht hervormer en verwoed verzamelaar'. In *Gids bij de tentoonstelling 'Het vergeten fenomeen. Sebald J. Brugmans, 1763–1819: Verzamelaar tussen koning, keizer en universiteit'*, edited by Tim Huisman, 9–45. Leiden: Museum Boerhaave, 2001.

Jonge, Hans de. 'Macht, machinaties en musea: Jan van der Hoeven, Hermann Schlegel en hun strijd om het Rijksmuseum van Natuurlijke Historie te Leiden'. *BMGN – Low Countries Historical Review* 120(2005): 177–206.

Klaauw, Cornelis J. van der. *Het hooger onderwijs in de zoölogie en zijne hulpmiddelen te Leiden: Een historische schets naar aanleiding van het 50-jarig bestaan van het tegenwoordige zoölogisch laboratorium*. Leiden: Sijthoff, 1926.

Klaauw, Cornelis J. van der. 'Een verzameling uit den eersten tijd van de vergelijkende ontleedkunde (de collectie Brugmans)'. *De Natuur* 50(1930): 49–53, 73–78, 97–101, 122–26.

Klep, Paul M. M., and Brand Kruithof. 'The Rise of Quantification and Statistics in Dutch Medical Research (1850–1940)'. In *The Statistical Mind in Modern Society: The Netherlands 1850–1940*. Vol. 2, *Statistics and Scientific Work*, edited by Ida H. Stamhuis, Paul M. M. Klep, and Jacques G. S. J. van Maarseveen, 11–37. Amsterdam: Aksant, 2008.

Koning, Pieter. *Beschrijving van Chineesche schedels*. Leiden: Van der Hoek, 1877.

Larson, Frances. 'Anthropological Landscaping: General Pitt Rivers, the Ashmolean, the University Museum and the Shaping of an Oxford Discipline'. *Journal of the History of Collections* 20(2008): 85–100. https://doi.org/10.1093/jhc/fhm020.

Matyssek, Angela. *Rudolf Virchow, das pathologische Museum: Geschichte einer wissenschaftlichen Sammlung um 1900*. Schriften aus dem Berliner Medizinhistorischen Museum 1. Darmstadt: Steinkopff, 2002.

Maulitz, Russell C. *Morbid Appearances: The Anatomy of Pathology in the Early Nineteenth Century*. Cambridge: Cambridge University Press, 2002. First published 1987.

Maulitz, Russell C. 'Pathology'. In *The Cambridge History of Science*. Vol. 6, *The Modern Biological and Earth Sciences*, edited by Peter J. Bowler and John V. Pickstone, 367–81. Cambridge: Cambridge University Press, 2009.

Molhuysen, P. C. *Bronnen tot de geschiedenis der Leidsche Universiteit*. 7 vols. The Hague: Nijhoff, 1913–1924.

Nyhart, Lynn K. *Biology Takes Form: Animal Morphology and the German Universities, 1800–1900*. Chicago: University of Chicago Press, 1995.

Outram, Dorinda. *Georges Cuvier: Vocation, Science, and Authority in Post-Revolutionary France*. Manchester: Manchester University Press, 1984.

Paget, James. *Descriptive Catalogue of the Pathological Specimens Contained in the Museum of the Royal College of Surgeons of England*. Vol. 4, *Morbid Conditions of the Urinary Organs, of the Nervous System and Organs of Special Senses, of the Generative Organs and Breast, and the Anatomy of the Stumps*. 2nd ed. London: Churchill, 1885.

Pickstone, John V. 'Museological Science? The Place of the Analytical Comparative in Nineteenth-Century Science, Technology and Medicine'. *History of Science* 32 (1994): 111–38.

Porter, Theodore M. *The Rise of Statistical Thinking, 1820–1900*. Princeton: Princeton University Press, 1986.

Rheinberger, Hans-Jörg. 'Präparate – "Bilder" ihrer selbst: Ein bildtheoretische Glosse'. In *Oberflächen der Theorie*, 9–19. Bildwelten des Wissens: Kunsthistorisches Jahrbuch für Bildkritik, vol 1:2. Berlin: Akademie Verlag, 2003.

Rheinberger, Hans-Jörg. *An Epistemology of the Concrete: Twentieth-Century Histories of Life*. Translated by G. M. Goshgarian. Durham: Duke University Press, 2010.

Rooy, Laurens de. *Snijburcht: Lodewijk Bolk en de bloei van de Nederlandse anatomie*. Amsterdam: Amsterdam University Press, 2011.

Royal College of Surgeons. *Descriptive and Illustrated Catalogue of the Physiological Series of Comparative Anatomy Contained in the Museum of the Royal College of Surgeons in London*. 5 vols. London: Taylor, 1833–1840.

Sandifort, Gerard. *Museum Anatomicum Academiae Lugduno-Batavae, Volumen tertium*. Leiden: Luchtmans, 1827.

Sandifort, Gerard. *Museum Anatomicum Academiae Lugduno-Batavae, Volumen quartum*. Leiden: Luchtmans, 1835.

Sandifort, Gerard. *Tabulae craniorum diversarum nationum*. 3 vols. Leiden: Luchtmans, 1838–1843.

Schickore, Jutta. *The Microscope and the Eye: A History of Reflections, 1740–1870.* Chicago: University of Chicago Press, 2007.

Swaving, Cornelis. 'Eerste bijdrage tot kennis der schedels van volken in den Indischen archipel'. *Natuurkundig tijdschrift voor Nederlandsch Indië* 23(1861): 241–89.

Sysling, Fenneke. 'The Archipelago of Difference: Physical Anthropology in the Netherlands East Indies, ca. 1890–1960'. PhD diss., VU University Amsterdam, 2013.

Sysling, Fenneke. '"Not Everything That Says Java Is from Java": Provenance and the Fate of Physical Anthropology Collections'. In *The Fate of Anatomical Collections*, edited by Rina Knoeff and Robert Zwijnenberg, 195–210. The History of Medicine in Context. Farnham: Ashgate, 2015.

Sysling, Fenneke. *Racial Science and Human Diversity in Colonial Indonesia.* Singapore: NUS Press, 2016.

Theunissen, Bert. *Nut en nog eens nut: Wetenschapsbeelden van Nederlandse natuuronderzoekers, 1800–1900.* Hilversum: Verloren, 2000.

Vanpaemel, Geert. 'Het getal regeert de wereld: Adolphe Quetelet en de wetten van de samenleving'. In *De opmars van deskundigen: Souffleurs van de samenleving*, edited by Frans van Lunteren, Bert Theunissen, and Rienk Vermij, 45–57. Amsterdam: Amsterdam University Press, 2002.

Virchow, Rudolf Ludwig Karl. *Das neue Pathologische Museum der Universität zu Berlin.* Berlin: Hirschwald, 1901.

Wallé, Dalila. *Leiden Medical Professors 1575–1940.* Leiden: Museum Boerhaave / Leids Universitair Medisch Centrum, 2007.

Whitehead, Christopher. *Museums and the Construction of Disciplines: Art and Archaeology in Nineteenth-Century Britain.* London: Duckworth, 2009.

Zaaijer, Teunis. *Beschrijving van twee vrouwenbekkens uit den Oost-Indischen archipel.* Leiden: Van Doesburgh, 1862.

3 Dead body in the closet

How lay visitors disappeared from institutional anatomical collections

A word of practical advice: never marry off your daughter to an old man she detests, however rich he might be, for you will be left with nothing but monstrous grandchildren. The proof of this rule-of-thumb could be found in the early nineteenth-century Leiden Anatomical Cabinet, where the product of such a marriage was on display (see Figure 3.1). The child, a boy, was the son of an exquisitely beautiful woman whose parents had forced her to marry a senile usurer. The man horrified the girl, but he was wealthy and therefore pleased the parents. The marriage was as short as it was unhappy: seven months after the ceremony, the woman and her baby died in child-birth. The baby did not look like a newborn child, but like an old man. And not just any old man – he was the perfect image of his father in miniature, down to the last wrinkle, as was explained on a tablet hanging next to the preparation that had been made of the boy.[1]

The tale on the tablet helped early nineteenth-century lay visitors to the Cabinet to make sense of the preparation. For them, it functioned as marriage advice; for the present-day viewer, it no longer does. Although doctors still think that older fathers increase the risk of malformed children, the preparation of the wrinkled boy can no longer be used to warn laypersons about this risk.[2] This has two reasons. First, the Leiden anatomical collections have become difficult to enter for those who are not (future) doctors. And second, even if you were to manage to get access to the preparation (now in storage), you would learn nothing about the boy's parents. No tablet, label, or guide is there to tell the tale, nor is it mentioned in the object description in the museum database. Not only has the preparation become practically unreachable for lay audiences, it has also been separated from the marriage story. Instead, the museum database describes the preparation in modern medical terms: 'anencephalus and rachischisis'.

The wrinkled boy's journey – from marriage advice exhibited in the Anatomical Cabinet to medical object stored away from public view – illustrates how it has become increasingly difficult for non-medical audiences to access anatomical collections. Many present-day institutional anatomy collections may be open to the public in principle, but are quite hard to enter in practice. Museum der Charité in Berlin, Barts Pathology Museum in

Figure 3.1 Preparation of a malformed child. According to an early nineteenth-century travel journal, the child was the son of a beautiful woman who had been forced to marry a senile usurer. Drawing by Abraham Delfos in Eduard Sandifort, *Museum Anatomicum Academiae Lugduno-Batavae, Volumen secundum* (Leiden: Luchtmans, 1793). Courtesy of Leiden University Library, PLANO 49 Y 2.

London, and Museum Vrolik in Amsterdam are examples of anatomical collections housed in medical institutions located outside city centres, which are thus more difficult to get to than the average art museum. Moreover, within these spaces, preparations are usually presented in a medical context – no stories about unhappy marriages to which the casual visitor can easily relate. This contrasts with the situation in the seventeenth, eighteenth, and early nineteenth centuries, when many anatomical collections were easily accessible to lay audiences – in Leiden, but also, for example, in Amsterdam, Rotterdam, Copenhagen, Altdorf, and Oxford.[3] The second half of this book argues that non-medical audiences lost access to anatomical collections because medical audiences continued to use the collections in medical research and teaching, as we saw in the first two chapters. Since the collections remained in use, they were affected by changing practices and attitudes in medicine, which led to new locations and different forms of presentation. To show how this shift hindered lay visitors, this chapter analyses the move and rearrangement of Leiden University's main anatomical collections, those in the Anatomical Cabinet. The next chapter investigates the consequences of the changes for another non-medical audience, institutional administrators.

This chapter first shows how the Leiden Anatomical Cabinet was a popular destination for travellers in the seventeenth, eighteenth, and early nineteenth centuries. It then discusses the transformation around 1860, when the Cabinet was moved to a new location, a laboratory complex away from the city centre. We will see how the move and the accompanying rearrangement made the collections both less approachable and less interpretable, and thus inaccessible to lay audiences.

The Anatomical Cabinet until 1860: open to all

Like all proper tourist destinations, mid-nineteenth-century Leiden had a beaten track. The Dutch author Nicolaas Beets (1814–1903) painted a lively image of this track in his *Camera Obscura* (1851):

> On this rainy October day, Hildebrand could be seen running through Leyden's streets together with a stranger, on their way to visit first the dead animals in the museum for natural history, and then the dead pharaohs in the museum for unknown history; and subsequently to take a look at Anatomy's little children who never lived, and then at the portraits of dead professors who will live forever in the senate hall … In order to establish some variety, we subsequently visited the Burcht [a fortress], which is a corpse itself, occupied by the Romans in earlier times, ADA, and the chamber of rhetoric to which so many geniuses belonged. To conclude, we went and saw Mr Siebold's Chinese and Japanese furniture, and finally we reposed at the student association building Minerva.[4]

Many of the sights mentioned by Beets were linked to the university: the Senate Hall and the Anatomical Cabinet, of course, but also the Museum for Natural History, the Museum of Antiquities ('the museum for unknown history'), and Minerva, the student association building. To its visitors, just as to its inhabitants, Leiden was first and foremost a university town, something that also shows from descriptions of the town's atmosphere. An anonymous British traveller (about whom we will learn more below) called it 'the most agreeable [town] to a person of studious habits' he had seen, because of 'a stillness, an absence of pomp, so congenial to academial pursuits, that you would rather fancy you were walking through the courts of an [sic] university, than through the streets of a populous city.'[5]

The university-related sights were all located close to one another, on or near Leiden's prettiest canal, the Rapenburg. At a bend in the canal, visitors could find the Anatomical Cabinet. The Cabinet was housed in the old Faliede Bagijnkerk (Church of the Faille-Mantled Beguines), which it had shared with the university library since the late sixteenth century. The combination of books and bodies was common back then. There were five universities in the early modern Netherlands; in four of them – Leiden, Groningen, Franeker, and Harderwijk – the anatomy department shared space with the library.[6] In the Bavarian town of Alt-dorf, the seventeenth-century anatomical theatre and library were housed next to each other, and anatomical objects could be found in both spaces.[7] The rationale for this was partly practical: a lack of space forced young universities to house diverse institutions together. But even as the universities grew, and more space became available, nobody felt the need to separate anatomy from the library – at least, not until the nineteenth century.

In Leiden, the library was closely connected not only with the anatomical collections and the anatomical theatre, but also with the botanical garden and its collection of rarities. Preparations, plants, and books went hand in hand: on one of the library's lists of acquisitions, an American crocodile pops up between the book titles.[8] The author of the list deemed it fit to include the crocodile because the animal belonged to the so-called book of nature – as did all animals, plants, body parts, and other natural objects. Nature was considered the second book of God, and studying it would, just like studying the Bible, bring one closer to God.[9] Both Catholics and Protestants employed the metaphor. As we read in the *Belydenisse des gheloofs* (Confessions of the faith, 1619 edition), one of the founding documents of the Dutch Reformed doctrine:

> We know Him by two means. Firstly by the creation, maintenance and reign of the whole world, since the world is before our eyes as a wondrous book, in which all creatures big and small are as letters which give us to behold the invisible things of God ... Secondly, He makes himself known even clearer and more fully by His holy and divine word.[10]

Anatomical collections (and other collections containing natural objects) were considered a chapter in the book of nature – not just in Leiden, but across Europe. The metaphor was used by many people in many places. Here is an example by Robert Hooke, curator of the London Royal Society's collections from 1662 to 1703:

> It were therefore much to be wishht [sic] for and indeavoured [sic] that there might be made and kept in some Repository as full and complete a Collection of all varieties of Natural Bodies as could be

obtained, where an Inquirer might ... peruse, and turn over, and spell, and read the Book of Nature, and observe the *Orthography, Etymolgia, Syntaxis,* and *Prosodia* of Nature's Grammar, and by which, as with a Dictionary, he might readily turn to and find the true Figure, Composition, Derivation, and Use of the Characters, Words, Phrases and Sentences of Nature written with indelible, and most exact, and most expressive Letters, without which Books it will be very difficult to be thoroughly a *Literatus* in the Language and Sense of Nature.[11] (italics in the original)

According to the book-of-nature metaphor, both nature and the Bible could be 'read'; both were the object of exegesis. Anatomists researching preparations and philologists analysing manuscripts engaged in the same activity: they were deciphering a text. Of course, their reading methods differed. Instead of literally reading the words, anatomists handled and dissected their texts – the book-of-nature metaphor does not contradict the hands-on use of anatomical preparations. The two types of books were read differently, but organized similarly. Both preparations and publications (as well as manuscripts) had to be described, classified, accessioned, placed, and catalogued. Together, the idea of the book of nature and the similar practices of organization meant that combining libraries and anatomy departments felt natural to early modern university governors, collection curators, researchers, and students.

For travellers to Leiden, the combination of university library and Anatomical Cabinet was convenient: they could visit two major sights in one building. Since the building stood in the town's centre, travellers could get there easily; and once they arrived, they had no trouble getting in. Figure 3.2 shows the building's entrance after the renovations of 1819–1822. You entered the Anatomical Cabinet through the door on the left; behind the door on the right were the stairs leading up to the library.

In 1850, both doors opened for attendees of the fifth Dutch rural-economical congress. At the request of the congress organizers, the university governors had asked all collection curators to give 'free access' to congress participants.[12] However, they failed to specify which kind of 'free' they meant: 'unrestricted' (as in 'free speech') or 'gratis' (as in 'free beer'). Anatomical curator Halbertsma, somewhat annoyed by the demand, wrote back and asked the governors to clarify:

I have to honour of letting Your Highly Esteemed Honourables know that the Museum Anatomicum is open to all and on every day. I call it 'free entrance' if a Cabinet can be visited by ringing at its door or by reporting to the custos, who lives right next to the building, and so I state that I do not understand what purpose the proof of attendance of the Rural-Economical Congress should serve.

Figure 3.2 Front entrance to the Anatomical Cabinet in the Faliede Bagijnkerk
(Church of the Faille-Mantled Beguines) after the 1819–1822 renovations.
The Cabinet was housed here until 1860. Detail of 'Plattegrond der stad
Leiden' (street map of the city of Leiden), lithograph by F. Desterbecq,
published by L. Springer, c. 1832. Courtesy of Leiden University Library,
Collection J. T. Bodel Nijenhuis, Port 14 N 52.

However, if the organizers of the above-mentioned Congress understand
'free entrance' as not paying 10 or 25 cents to the custos, I feel obliged
to stand up for his interests. Tips from visitors to the Museum Anatomicum form a substantial part of his income, and hence it would be an
unpleasant disappointment if they were to be withheld from him on this
occasion, especially when one realizes that the congress participants will
not hesitate to spend considerably larger sums of money on less scientific purposes during the three days of the conference.[13]

Halbertsma suggested that a box be placed at the entrance of the Cabinet, so
that each congress visitor could donate a small amount. But within a few
days he had withdrawn his proposal – as if it were a hasty, angry e-mail –
and asked the governors to act as though they had never received his letter.
We do not know how (and if) the governors reacted. Whatever eventually
happened during the rural-economical congress, though, Halbertsma's letter
does reveal the daily routine: the Anatomical Cabinet was open to all, at a
small cost. The opening hours were long; Halbertsma writes that it was
open 'every day'. (It is unclear whether this included Sundays; according to
the student almanacs, the Cabinet was closed on Sundays.) During opening

hours, one could gain access by simply ringing the bell, or, if nobody answered, by knocking on the door of the neighbouring house where the custodian (*custos* in contemporary terms; see Halbertsma's letter) lived. Recommendation letters and prior arrangements were unnecessary; as Halbertsma stated in his letter, the Cabinet was open to all. It always had been: from their foundation in the late sixteenth century onwards, the Leiden anatomical collections had been a major tourist attraction, easy to access.[14]

Medical historian Rina Knoeff has described the early modern Leiden anatomical collections as 'visitable', a notion borrowed from Bella Dicks.[15] A visitable place is, as Dicks puts it, 'somewhere to go'.[16] It is a *destination* – and that is indeed what the old Cabinet was. To become a destination, or to be visitable, a collection needs to be accessible in more than one sense. It needs to be both *approachable* and *interpretable*. An approachable collection is a collection that is easy to enter, like the pre-1860 Leiden anatomical collections were. As we will see, they were also interpretable, which means that visitors could easily engage with them and make sense of them. I chose the word 'interpretable' to denote this kind of accessibility because it indicates visitor agency more clearly than, for example, 'intelligible'. Visitors did not just passively take in what they were told; they actively constructed their own interpretations.

One such visitor was an anonymous British military man, who wrote about the Cabinet in one of his letters home. The letters were later published under the title *Billets in the Low Countries, 1814–1817*. He recalled the above-mentioned story about the monstrous child of the beautiful woman and the old usurer. Moreover, he added his own experience with the preparation in the Cabinet, and his account shows that he was both physically and emotionally close to the preparation.

Writing about the monstrous child, the military man tells us that 'by means of a glass you can trace every wrinkle, and verify every property of age'.[17] Apparently, visitors were invited to come close and engage with preparations, in this case to verify for themselves that the wrinkled boy indeed possessed all the characteristics of an elderly man. Thus, visitors came physically close to the preparations, albeit perhaps not as close as researchers and students, who could remove such preparations from their jars. We do not know whether lay visitors were allowed to handle preparations in the way students and researchers did. I have found no direct evidence of such handling, but it is not unthinkable: it happened earlier, and elsewhere. Rina Knoeff has argued that visitors may have been allowed to touch and hold anatomical preparations in the seventeenth-century cabinet of the Amsterdam anatomist Frederik Ruysch.[18] A nineteenth-century example of laypersons handling preparations can be found in mid-nineteenth-century Vienna. Here, comparative anatomy professor Carl Brühl lectured to a broad audience, including many women. Brühl let them handle preparations, as the following reports from the *Wiener Medizinische Wochenschrift* demonstrate:

Some of the ladies, who until now had been satisfied only with the finest perfumes, heroically ignored completely the alcoholic stench of a brain of a fellow human being hardened in the strongest alcohol, to be able to scrutinize its complex surface more accurately with their own delicate fingers.[19]

And, a year earlier: 'At last the most delicate ladies held the human brain parts in their hands as courageously as any medical student.'[20]

Collection visitors are not passive recipients of information; they actively interpret what they see (and touch, and smell, and hear). They add their own knowledge and experience to the presented objects – a practice Samuel Alberti has called 'the museum affect'.[21] The author of *Billets*, for example, first described the preparation of the monstrous child, then told the story of the marriage, and finally reflected upon this story and the preparation, creating his own interpretation:

> This corporeal resemblance of the father, in the shape of this little prodigy, seems to have been flung upon the world by indignant nature to shame those who would defeat her purposes by a rebellious opposition to her laws. ... It would certainly serve as a clue to ascertain why matrimony is so often the source of misery. Some blame fortune, others destiny; but all forget the share which policy has in the contrivance.[22]

The author used his ideas on nature and marriage to make sense of the preparation. But he was only able to do so because he had been offered the story about the parents of the monstrous child. That story enabled him to engage with the preparation not just physically (by looking at it closely), but also emotionally.

Early modern visitors to the Leiden collections engaged with the preparations in similar ways to the author of *Billets*. They interacted with the preparations both physically and emotionally, but they were only able to do so because of the stories provided by the collection's catalogue and tour guides.[23] The stories made the preparations interpretable. Take, for example, the skeletons in the anatomical theatre. Without context, skeletons were not the most interesting preparations – they could be seen everywhere, and they all looked alike. Visitors needed a point of departure to interpret each skeleton individually. In Leiden, the skeletons were presented with reference to the crimes committed by the people they had once been. These crimes were even narrated in the collection's catalogue, which listed (in what to our modern eyes seems a somewhat idiosyncratic spelling), for example, 'the Sceleton of an Asse upon which sit's a Womam that Killed her Daughter'; 'the Sceleton of a Man, sitting upon an ox executed for Stealling of Cattle'; and 'a young thief hanged being the Bridegom whose Bride stood under the gallows, very curiously set up in his ligiments'.[24] The crimes individualized the skeletons. Furthermore, many of the Leiden skeletons carried banners showing Latin phrases such as *Nascentes morimur* (We are born but to die), *Nosce te ipsum* (Know thyself), and *Mors ultima linea rerum* (Death is the final limit of all things). In this context, it became possible for visitors to interpret the otherwise very similar (and rather boring) skeletons in an individual and exciting way.

In short, from the late sixteenth to the early nineteenth century, the Leiden anatomical collections were both approachable and interpretable: visitors could easily get into the building, they could get physically close to the preparations, and they could relate to the preparations emotionally and intellectually. Although lay visitors had little knowledge of anatomy, they had no trouble making sense of the preparations (Figure 3.3).

The early modern Anatomical Cabinet was remarkably accessible compared to other types of collections at the time. In his canonical book *The Birth of the Museum*, sociologist and museum studies scholar Tony Bennett describes early modern collections as 'socially enclosed spaces to which access was remarkably restricted'.[25] The description 'remarkably restricted' in no way applies to the early modern Anatomical Cabinet. This is partly because Bennett is writing about European collections in general and British

Figure 3.3 In the early modern Leiden anatomical theatre, the anatomical collections appealed to lay visitors. Anonymous etching, 1712. Courtesy of Rijksmuseum Amsterdam, RP-P-AO-10-27C-2.

collections in particular. Understandably, he pays no attention to the spe-cifics of the Dutch situation, which seems to have been quite different: in the Dutch Republic, most types of collections were more open than the ones Bennett describes.[26] But even by Dutch standards, the Anatomical Cabinet was unusually open. Many of the art collections in the Republic were pri-vately owned and open to a select audience only,[27] and collections (art or otherwise) accessible to wider audiences often had more limited opening hours than the Anatomical Cabinet. In 1774, stadtholder William V opened his collections to the public, but not every day, and only between eleven and one o'clock.[28] Furthermore, gaining access was often more difficult than simply ringing the bell: in Teylers Museum (founded in 1784), for example, every visitor required prior approval from the board of trustees.[29]

Compared to other *anatomical* collections, however, the Leiden Cabinet was less exceptional. Many other anatomical collections were also open to a broad audience. In the Dutch Republic, accessible anatomical collections (usually housed in anatomy theatres) could also be found in Amsterdam, Delft, Dordrecht, Rotterdam, Utrecht, Franeker, and Middelburg.[30] Else-where, laypersons could come and see anatomical objects, for example, in the Jardin du Roi in Paris and at the universities of Copenhagen, Altdorf, and Oxford.[31] Just as in Leiden, the preparations were presented in a way that appealed to lay visitors. In Oxford, visitors learned that one of the skeletons had belonged to a woman who had had 18 husbands and had murdered 4 of them.[32] In Altdorf, the skeleton of a Croatian criminal could be seen holding a spear, sitting on his horse.[33]

The accessibility of these collections seems remarkable in the history of museums and collections, but it makes perfect sense in the history of anatomy. Long before universities started building significant anatomical collections, the discipline of anatomy was already welcoming non-medical audiences, at its public dissections. The first known European public dissection took place in 1316 – almost 300 years before Pieter Pauw acquired some bones and began the Leiden collections, and approximately 350 years before anatomists developed techniques to create long-lasting fluid preparations. Public dissections attrac-ted a mixed crowd: not just physicians, surgeons, and medical students, but also laypersons, including many dignitaries. The non-medical spectators had no trouble understanding what was going on, because the public dissection was not so much a medical event as a religious ritual and a moral-philoso-phical lesson.[34] The audience was meant to marvel at the make-up of the human body, the Creator's masterpiece. In other words: visitors to a public dissection were reading a chapter from the book of nature. The visitors were also participating in a ritualistic public punishment. Often, the body on the table (had) belonged to a convicted criminal; public dissection after death was considered an additional punishment.[35] The strong religious and moral message of public dissections made them comprehensible and appealing to non-medical audiences. In a similar way, early modern anatomical prepara-tions were not exclusively about bodily structures, but also about the workings of the soul, about morality, and about biblical lessons – things that mattered to wider audiences than medical students and professors alone. Early modern anatomy was public, morally loaded, and religious, and thus it is not surprising that the anatomical collections of the age were easily accessible to a wide range of audiences.

In Leiden, as we will see below, visitors ceased to visit the Cabinet from 1860 onwards. British anatomical collections were also closed off around this time.[36] The loss of accessibility was caused by changes in anatomy, as well as in medicine as a whole. The discipline of anatomy started to change fundamentally towards the end of the eighteenth century – and with these changes, it lost its public, religious character. In Leiden, the public anatomy theatre was demolished during the renovations of 1819–1822. Anatomical theatres were dismantled everywhere: medical historian Andrew Cunningham has shown that public dissections disappeared throughout Europe between 1780 and 1830.[37] Cunningham has linked this disappearance to the secularization of the world view in general and of the natural sciences in particular, to the disappearance of other types of public events (such as the public execution), and to the rise of expertise – developments that occurred simultaneously, although the rise of expertise, as we will see below, reached its peak somewhat later, in the second half of the nineteenth century (and continued well into the twentieth century). In the first half of the nineteenth century, a new kind of anatomy emerged, which was neither religious nor public.[38] And it was not just anatomy that changed. As the nineteenth century unfolded, medicine as a whole was transformed. Researchers used new methods and new theories, which were increasingly based on physics and chemistry. As we saw in Chapter 2, the new medicine continued to use the old collections. Eighteenth-century preparations were reinterpreted using nineteenth-century techniques such as microscopy and statistics. With these reinterpretations, preparations continued to function in teaching as well. As a result, researchers and students took the preparations with them when, in the second half of the nineteenth century, they moved to a new space that would come to characterize modern medicine: the laboratory. After 1850, laboratories became essential in both medicine and the natural and life sciences; first in teaching, then in research.[39] As we have seen, contrary to common opinion, laboratories did not exclude collections. They did transform them, however, and through this transformation, the collections lost their accessibility to lay visitors. To understand how this happened, let us now examine the Leiden Anatomical Cabinet's move from the library to the laboratory.

1860: from the library to the laboratory

Anatomical collections were moved into laboratories across Europe in the second half of the nineteenth century, but the exact timing of each move depended on specific local circumstances. In Leiden, the process began with a sagging ceiling, that of the Cabinet's collection room. The ceiling doubled as the floor of the university library, a fact no-one took much notice of until the early 1850s, when this construction started to create problems. The ceiling began to sag under the weight of the library's books. Two iron pillars prevented a collapse, but the situation was less than ideal.[40] And, as though imminent collapse were not enough, curator and professor Hidde Halbertsma faced more architectural problems. The Cabinet was unfit for teaching experimental physiology. Halbertsma was responsible for the physiology course, as he held the chair in anatomy and physiology, which would not be divided into two chairs until after Halbertsma's death in 1865. In his 1851–1852 annual report, Halbertsma elaborated upon one of the problems he encountered in teaching physiology:

> At the moment, both lecture rooms available to me are amphitheatric [with the students seated in a semi-circle] and they can therefore be considered less suitable for physiology lectures. With the present layout, listeners at the front regularly turn their backs on the Professor, which, in my opinion, cannot have a particularly positive effect on their attention, especially because more difficult subjects have to be clarified with the aid of hand-made drawings on the blackboard.[41]

Halbertsma implied that the problem of students looking the other way did not arise in anatomy lectures; unfortunately, he did not explain why. Perhaps it had to do with the nature of physiological experiments.[42] Physiology lectures involved both demonstrations and drawings on the blackboard to explain the experiments. Unlike anatomical demonstrations, physiological experiments could not easily be interrupted and continued, meaning that the students had to look at the blackboard, the demonstration table, and Halbertsma at the same time. The amphitheatric layout may have prevented them from doing so, for example, if the demonstration table stood inside the semi-circle of seated students and the blackboard was positioned more to the side, (almost) outside the semi-circle. Although this is not known for sure, we do know that Halbertsma claimed that he lacked a decent classroom for his physiology lectures. Furthermore, the Anatomical Cabinet did not contain a teaching laboratory, which also was essential for physiology teaching, as Halbertsma stated repeatedly in his annual reports.[43]

Neither the amphitheatric arrangement in the lecture rooms, nor the absence of a physiological teaching laboratory had bothered Halbertsma's predecessor, Gerard Sandifort. And yet Sandifort, like Halbertsma, taught both anatomy and physiology. He did so in a completely different way, however, as is illustrated by the course descriptions in the *series lectionum*. Sandifort's course was described as 'Physiologiam, anatome comparata illustratam' ('Physiology illustrated through comparative anatomy'); Halbertsma's as 'Physiologiam, experimentis et observationibus microscopicis illustratam' ('Physiology illustrated through experiments and microscopic observations').[44] Here we see the difference between what Andrew Cunningham has called 'old' and 'new' physiology.[45] Old, early modern physiology was a theoretical, philosophical discipline based on the study of form and best taught through Latin lectures illustrated with anatomical material. New physiology, which emerged in the first half of the nineteenth century and arrived in Leiden around 1850, was an experimental discipline which explained the working of the body through physical and chemical processes instead of morphology, and which was best taught through a combination of lectures and practical training in microscopic observations and (animal) experiments. Sandifort taught old physiology and thus required nothing more than an amphitheatric lecture room. Halbertsma, on the other hand, taught new physiology and hence required a lecture room with a blackboard on which he could draw the chemical and physical processes in the body, a room where students could practise with microscopes, and a teaching laboratory where students could perform experiments themselves.

Halbertsma was not the only Leiden professor dissatisfied with his teaching facilities. Professors Petrus Rijke (physics) and Anthony van der Boon Mesch (chemistry) also complained to the governors.[46] As in medicine, teaching laboratories became more and more important in physics and chemistry. (In fact, the teaching laboratories in the natural sciences served as an example for educational reformers in medicine.) Both departments had spaces for practical training, but these were ill-equipped and too small. Rijke and Van der Boon Mesch each repeatedly asked for new laboratories from 1846 onwards. Van der Boon Mesch was backed up by his students (in 1851 and 1852) and by a group of Leiden citizens, including several industrialists (in 1851). At first, the governors refused the professors' requests, but after several years, they yielded.[47] To solve all the problems at once, they planned a new building to house physics, chemistry, and anatomy. Anatomy would be separated from the library and merged with the natural sciences. This shift agreed with the changes that the discipline of anatomy had undergone: the religious book-of-nature metaphor had lost ground, and physics and chemistry had become ever more important in its practice.

The governors chose the Ruïne (the Ruins) as the location for the new building. In 1807, an exploding gunpowder ship had destroyed all the buildings in this area. The university had drawn up its first plans to build on this spot soon after, but none of them had been implemented, although the first stone for a commemorative column had been placed.[48] In 1854, the university governors sent their new proposal to the responsible minister. The minister agreed on the need for a new building, but rejected the governors' plan due to the estimated cost of 200,000 guilders. He asked the government architect Henri Camp to create a new, cheaper design. In 1857, the Utrecht contractor Van Berkum drove the first pile into the ground, and the building was completed some two years, and several financial setbacks, later.[49] In 1859, the physics and chemistry departments moved in, followed by anatomy in 1860 (Figure 3.4).

Halbertsma was pleased with the Anatomical Cabinet's new home. In his first annual report following the move, he wrote:

Figure 3.4 Front entrance of the new teaching complex for physics, chemistry, and anatomy. Visitors to the anatomy department had to enter at the back of the building; the fence enclosed the entire area. Photograph by Ad. Braun and J. Goedeljee, 1866. Courtesy of Erfgoed Leiden en Omstreken, Leiden, PV10.45.

Although not everything in the present complex meets the demands that we believe to be justified, for now, we are glad about the major improvements that have resulted from the move. These improvements concern in particular the lecture rooms, the dissection hall, the work-rooms, the arrangement of the cupboards, [and] the lighting, not to mention many other things, which are out of place in a report like this, and which I discussed in more detail when I had the honour of inaugurating the academic year on the new premises on October 1st, 1860.[50]

Halbertsma admitted that the new building was not perfect – for example, it would take until 1866 before a proper physiological laboratory was added to the site – but all in all, he thought it much better than the old one.

Not everybody shared Halbertsma's approval, though. When describing the building, the student almanac of 1860 posed the following rhetorical question:

This building as it is seen from the outside, with its humble façade, with its ridiculous, ambiguously spherical back section, with its little garden divided into four beds, with its wooden fences – should we not call it, from an architectonic point of view, a *monstrum horrible visu*?[51]

The students not only criticized the architecture; they also judged the anatomy section to be too small.[52] Indeed, a few years later, an additional gallery had to be added to one of the collection rooms to accommodate the newly acquired Suringar collection.[53] Not long after that, in the 1870s, lack of space once again became a problem: the annual report of 1883–1884 stated that some of the students 'had to seat themselves on the stairs and even on the edge of the sink' and that many could not see the demonstrations.[54] Several extensions were added in the 1880s to accommodate the growing anatomy department; meanwhile, physics professor Kamerlingh Onnes was slowly taking over the main building.[55]

Another audience that probably had mixed feelings about the Cabinet's new location was that of lay visitors. Unlike the students, they did not explicitly voice their concerns, which is not surprising, considering that they were a far more heterogeneous and far less (or rather, not at all) organized group. Instead of criticizing the new space in writing, visitors voted with their feet: after the move, visitor numbers seem to have dropped sharply. Unfortunately, this decrease is impossible to prove with hard figures. The only quantitative records that we have are from the period after 1860, and their accuracy is questionable. Nonetheless, for several reasons it is safe to assume that the Anatomical Cabinet was visited much less after it was moved from the library to the laboratory.

Let us take a closer look at the figures we do have: the name counts from the only known visitor book of the Anatomical Cabinet, which starts in September 1860, directly after the move.[56] Visitor books provide some insight into who came to see a collection, but it is hard to estimate what percentage of visitors actually signed the books; by no means every visitor signed his or her name. Consider, for example, the register of visitors kept between 1805 and 1932 at the Royal College of Surgeons in London. It lists fewer than a hundred names for the entire nineteenth century, whereas other sources reveal that the period between 1815 and 1830 alone saw over 25,000 visitors – and for this collection, the annual number of visitors would only rise as the century progressed.[57] Why they were not represented in this register becomes immediately clear upon reading its name: 'register of illustrious and distinguished visitors'.[58] Only the most important visitors were allowed to sign it: page after page, it lists the names of princes, dukes, bishops, and ambassadors. The register served to enhance the collection's status, not to meticulously record its visitors. This type of visitor book was not uncommon at the time, but other, more inclusive ones were used as well. However, these latter registers were not always more representative, as is shown by the visitor books at the Rijksmuseum in Amsterdam. In 1879, 36,218 people visited the Rijksmuseum in Amsterdam, but only 2,923 of them are listed in the visitor book.[59] The problem was not that people were not allowed to sign, but that they were not obliged to – and, as you may know from experience, many people simply walk right by.

The Cabinet's visitor book was probably not very exclusive, as it was signed by a range of visitors, both Dutch and foreign, doctors and non-doctors, the latter including Leiden professors from other faculties and several members of Halbertsma's family. On 21 March 1861, for example, the book was signed by the Leiden professor in history Robert Fruin. Fruin signed without his title, as did others: most signatures lack a title, even if the visitor, like Fruin, did possess one. This is another indication that the book was not initially intended as a status symbol; it appears all those who visited the Cabinet could sign their name. Nonetheless, the number of visitors listed is limited. In the early 1860s, between 20 and 40 people visited each year (with a peak of 84 visitors in 1863). From 1865, numbers dropped to an average of four visitors a year. Finally, after 1876, no more names were added, although the book still held 203 empty pages. These are negligible numbers compared to those in the visitor books of other collections at the time – recall, for example, the almost 3,000 signatures in the Rijksmuseum visitor book. There is no reason to assume that visitors were less inclined to sign the visitor book at the Cabinet than they were when visiting collections elsewhere. Hence, we can assume that visitor numbers in the Cabinet were low compared to other collections at the time.

Furthermore, if a visitor book had been kept before 1860, it would have contained more names – even if only a small number of visitors signed it. Although we have no visitor numbers, we can roughly estimate the order of magnitude with the aid of data that we do have: the number of visitors to one of the other Leiden collections, the Museum of Antiquities. This museum opened in 1838 and received 3,000 visitors in its first year.[60] Since travel journals from this period indicate that the Anatomical Cabinet was one of the main attractions in Leiden, we can safely assume that its visitor numbers resembled those of the Museum of Antiquities. This means that it is not unlikely that the Cabinet received thousands of visitors each year, which equals around a dozen a day. Even if only 1 percent of these visitors signed a visitor book, it would contain ten to a hundred times as many names as the visitor book starting in 1860. This means that visitor numbers after the move were low, not only compared to other contemporary collections, but also compared to the old Cabinet. Laypersons were no longer visiting the collections.

Elsewhere, institutional anatomical collections were often just as inaccessible as in Leiden. Although British anatomical collections started targeting a broader audience in the 1830s and 1840s, they began to discourage or even refuse lay visitors soon after.[61] The Anatomy Museum of the University of Basel accepted lay visitors on Sunday afternoons only, when they could view just a small part of the collections.[62]

The closing off of institutional anatomical collections contrasts with the nineteenth-century rise of what Bennett has called the 'exhibitionary complex', in which more and more collections became publicly accessible.[63] This new exhibitionary complex included popular anatomical museums, by which I mean not just anatomical collections open to a wide audience (like the early modern Leiden collections), but a specific, nineteenth-century kind of anatomy museum. Popular anatomical museums emerged in the 1830s and the 1840s (both in Europe and the United States); they were a commercial enterprise aimed at a broad, non-medical audience; and they displayed both wax models and preparations of the human body.[64] In England, the medical profession campaigned fiercely against the popular museums and exhibitions, and succeeded in shutting down most of them with the help of the Obscene Publications Act (1859). It was not hard to build an obscenity case against a popular anatomical museum – sex and crime were well-represented – but the most pressing concerns of the English medical professionals probably related less to morality than to a potential loss of income and a wish to monopolize medical knowledge.[65] In England, the owners of medical museums cooperated with quack doctors from the 1850s onwards to try and sell visitors as many cures (effective or not) as possible, and this directly threatened the medical profession. In the Low Countries, popular museums and exhibitions never started selling pharmaceuticals, which may partly explain why the medical profession there reacted somewhat less stridently.[66]

Leiden never had a permanent popular anatomical museum, but the town was visited by travelling exhibitions. Local newspapers announced them, for example the *Leidsch Dagblad* in April 1885:

> On the Bloemmarkt ['flower market', a street in Leiden] in this town, a tent is being built for the Anatomical Museum of Dr P. Spitzner from Paris. The museum contains 6,000 wax objects, representing complete bodies, human body parts, pathologies, etc. To judge from its extensiveness, the collection will exceed many others of its kind, well-known to us from fairs, in importance. The low entrance fee will certainly tempt many to come and see the collection. The museum will be open for a few days only, starting this Tuesday.[67]

The phrase 'well-known to us from fairs' reveals that Leiden regularly hosted popular anatomical exhibitions at this time. Most major towns in the Netherlands and Belgium hosted a travelling anatomical museum at least once a year, often on fairgrounds, especially from the 1850s onwards.[68] The travelling museums came mainly from France and the German-speaking lands. As stated in the announcement above, the Spitzner museum came from Paris, from where it travelled to fairgrounds in France, Germany, England, Belgium, and the Netherlands.[69] Although the local Leiden newspaper considered the size of the Spitzner collection remarkable, the type of collection was familiar. The success of popular exhibitions, both in Leiden and elsewhere, demonstrates that lay visitors did not stop coming to the Leiden Anatomical Cabinet because they had lost interest in (representations of) the human body. They still wanted to see anatomical objects, but they preferred popular anatomical collections to the Cabinet and other institutional collections.

Visitors were not actively refused entrance to the new Cabinet; lay people were still allowed to visit the collections, as the visitor book shows. However, being open to the general public does not in itself turn a place into a destination: it is necessary, but not sufficient. A visitable collection requires more: the building needs to be approachable and the objects inside need to be interpretable. Popular anatomical museums and exhibitions met these requirements – they had to, in order to make a profit. Until around 1850, the Anatomical Cabinet had met them as well, but in the second half of the nineteenth century, the Cabinet lost both its approachability and its interpretability. The remainder of this chapter explains how this happened.

A less approachable building

Visitors who wanted to enter the new Anatomical Cabinet had to overcome several hurdles. First of all, they had to walk a little further than they had done in the past. Before the move, the collections had been located in the centre of Leiden, close to other major sights; the academy building and the botanical garden could be found across the canal. The laboratory complex was situated somewhat further away from the town centre, with few other attractions nearby, let alone in the same building, as had been the case with the library. A longer walk was not insurmountable, of course, but it did pose a barrier to visiting.

Moreover, visitors encountered several challenges upon arrival at the Ruïne. In particular, they had to get to the entrance – which was less trivial than it sounds. Even the Cabinet's personnel struggled with this from time to time, as Halbertsma explained to the governors in 1861:

> Among the things urgently needing improvement in the new building at the Ruïne (anatomy department) are in the first place the entrances. These are faulty, both at the front and at the back, and hence, from time to time, the personnel belonging to my department have to cross the grounds of the wings or climb over the fence in order to get inside.[70]

The building area was enclosed by a fence with four gates, but it seems that the gate leading to the anatomy department did not always open easily, forcing Halbertsma's employees – and potential visitors – to put in some extra effort. Although the fence was not necessarily high, it made visiting the collections that much more difficult. And before visitors could discover that the anatomy gate stuck, they had to locate it. Finding the front gate was easy enough, but this gate was exclusively intended to access the physics and chemistry laboratories (although Halbertsma's staff sometimes used it as well, if all else failed). The Anatomical Cabinet was located at the rear of the building, or, as the student almanac put it, the 'ridiculous, ambiguously spherical back section', which meant that visitors had to find their way around the

building into the Zonneveldsteeg (Zonneveld alley).[71] Again, not insurmountable, but the back door was less welcoming than the front
entrance, especially when it was raining. Halbertsma again:

> At the back of the anatomical cabinet, by the gate leading to the Zon-
> neveldsteeg, is a small street, which is separated from the main street by
> a wide strip of soil, covered with coarse sand. After heavy rain, large
> puddles of water remain in front of this small street, which makes it
> impossible to enter the garden behind the Anatomical Cabinet properly
> through the gate.[72]

All in all, finding one's way in was much harder than it had been in the
Faliede Bagijnkerk. For more than two centuries, visitors had simply
entered the Anatomical Cabinet through a clearly recognizable front
entrance, facing Leiden's main canal. Now, they had to find their way to
the back alley, wade through the puddles, pray that the gate would open
(or climb over the fence), walk up to the building, and knock on the door.
If the custodian failed to answer they had to turn around, tackle the fence
and puddles again, find the custodian's house in the Zonneveldsteeg, and
hope that the gate would still open when they returned. But the difficulties
did not end there: even if visitors did manage to enter the building, it was
hard to find the collections. These were located in four rooms on the top
floor, instead of in the main room on the ground floor, as had been the
case in the old Cabinet.[73]

In response to Halbertsma's complaints, the situation improved somewhat: the governors ordered the inspector of the university's buildings to fix
the gates and asked the City of Leiden to pave the gap between the alley and
the gate.[74] But the collections remained hard to approach in the new building, just like other anatomical collections that were moved to teaching and
research laboratories in the second half of the nineteenth century. Although
broken gates and puddles were specific Leiden problems, the distant location
of the laboratory was not: since laboratories needed new, large buildings, in
most cities they were built on the outskirts of town – usually, there simply
was no space in the city centre. Moreover, wherever they were located,
laboratories were not inviting places for lay persons to visit.

A laboratory is a 'closed space'. This is reflected in its architecture (it was
no coincidence that the Leiden laboratory complex was fenced in), but also
in its atmosphere. A laboratory, whether for teaching or for research, is a
strictly regulated environment with clear target audiences: students and
researchers. Even if other groups are allowed in (which often they are not),
lay persons will, in general, hesitate to enter a laboratory. The strict and
many regulations – do not touch this, do not use that, wear a white coat –
create an intimidating atmosphere that scares off most potential visitors.
Even if the rooms housing the collections in the newly built laboratories did
not look 'laboratory-like', they were nevertheless located in buildings that

were usually known first and foremost as laboratory buildings, and as such had a closed atmosphere. In Leiden, the closed atmosphere became more prominent towards the end of the century, as the building on the Ruïne increasingly became a research laboratory. Again, this contrasted with the old Cabinet. Here, the collections had been housed in and around an anatomy theatre (until 1819), together with the library, in a church. All three spaces had open atmospheres: the theatre as the location of public dissections; the library as a tourist attraction; the church as the house of God. These open atmospheres reinforced each other, as well as the open character of the anatomical collections.

The relatively remote location and the closed atmosphere made anatomical collections housed in nineteenth-century teaching and research laboratories less accessible than their predecessors. In Leiden, the sticking gate and the puddles formed additional obstacles, but if a visitor did manage to reach the entrance of the anatomy building at the Ruïne, he or she would be let in. Leiden's curators did not explicitly reject laypersons. But few people took the trouble, because once they did get in, they were confronted with collections that no longer appealed to them. After the move, the collections had been rearranged and, as we will now see, in the new arrangement the preparations were difficult to interpret without prior medical knowledge.

A less interpretable arrangement

Although Halbertsma did not turn away visitors, he did keep some preparations away from them. He felt ashamed of the condition of the preparations:

> I can say the same [i.e., being in need of new fluid] of many preparations which are already listed in the Catalogue of the Museum Anatomicum, and hence were already present when I arrived here; they have been taken off the shelves for now, to avoid nasty and critical looks, and now they stand in a hidden corner, thirsty.[75]

After his appointment in 1848, Halbertsma discovered that many of the preparations were in bad shape. Many of the wet preparations had dried out; most of the skeletons were suffering from damp.[76] And it was not only the state of the individual preparations that bothered Halbertsma; he was also dissatisfied with the composition, classification, and arrangement of the collections as a whole. Determined to solve these problems, Halbertsma asked the governors for extra money and set to work together with his newly appointed prosector, Johannes Boogaard. By the mid-1850s, they had topped up the fluids, relabelled the jars, cleaned the skeletons, and varnished the bones.[77] They decided to wait a few more years before rearranging the collection, as this would require additional time and money, and Halbertsma thus preferred to wait until the move to the new building, for which the first plans had emerged.[78]

In the old system, the preparations had been arranged, by and large, by their makers. Halbertsma instead proposed to classify them systematically, by separating general anatomy, pathology, and comparative anatomy, and then organizing the objects according to organ system within these categories. He intended to follow the system used at the Royal College of Surgeons in London, the catalogues of which he acquired in the academic year 1854–1855 with the help of university governor D. T. Gevers van Endegeest and the Dutch ambassador in Britain.[79] The new classification system was put into operation after the move. Preparations deemed irrelevant in the new system were discarded; the remaining ones were put in their proper places on the shelves. Describing the preparations anew was also part of the job, but with thousands of preparations and little time at hand, it would take over 30 years and another two curators before this task would be more or less completed.

Halbertsma made all these changes with a clear aim in mind: he wanted collections fit for research and teaching. After a visit to the Anatomical Cabinet, the university Senate summarized Halbertsma's intentions:

> The director [of the Anatomical Cabinet, i.e., Halbertsma] is always inspecting and repairing the existing preparations, and separating the ones without use. … Rightly, with regard to extending the collection, it is not so much his intention to give the cabinet an appearance that amazes the general public or less experienced visitors because of its curiosities, but rather to possess a collection of objects useful and indispensable for teaching and research.[80]

Halbertsma considered it impossible to reach out to students, researchers, and lay visitors simultaneously, and he chose the first two audiences over the third. This distinguished him from his predecessors. In the early modern period, the Leiden anatomical collections had catered to students, researchers, and lay visitors combined. Preparations were presented in such a way that lay visitors could easily relate to them, but this did not make the collections unsuitable for research and teaching. The religious and moral issues that appealed to non-medical audiences formed an integral part of the discipline of anatomy. Of course, anatomists also investigated more specialist questions on bodily structures and functions. They used anatomical collections for these investigations as well, and although this use did not add to the accessibility of the collections to a wider audience, it did not threaten it either – the different uses simply co-existed.

As mentioned above, religion and morality disappeared from anatomy after 1800. Yet the Cabinet's first nineteenth-century curator, Gerard Sandifort, continued the early modern exhibition practices. It was during his period of appointment that the anonymous English visitor read the tablet on the unhappy marriage and traced the wrinkles on the monstrous child. Other travellers who visited the Cabinet in Sandifort's day mentioned similar interpretable preparations in their reports. Around 1805, for example, the American chemist Benjamin Silliman was shown a monstrous birth preserved in a large glass jar, visited annually by its mother for the last 19 years.[81] Jean Duchesne, who visited in the 1830s, wrote about the head of a giant called Cajanus.[82] Not only could Cajanus's head be seen, but also some of his clothes. We know this because another traveller, Karel van Wildenstein, felt the need to tell his readers that Cajanus's slipper was absent during his visit, as was 'the shoe of the infamous farmer of [the town of] Lekkerkerk'.[83] We would not recognize slippers and shoes as anatomical objects, nor would nineteenth-century anatomists. Rather than demonstrating facts about the human body, the footwear made the collections more interpretable to lay visitors. As did the fact that Cajanus had a name, and was not just one of many giants, but a unique individual – with his own slippers, which also helped visitors imagine how huge Cajanus's feet must have been.

Sandifort refrained from changing the composition of the collections or the descriptions of the preparations, because he was satisfied with the collections as they were. In his annual reports, he described the collections as rich and the condition of the preparations as good, and he never complained about the facilities; he requested additional money only for the publication of catalogues, so that the collection could be shared with a wider audience. An example from the 1837 report:

> The anatomical-physiological-pathological cabinet ... has already acquired such an extensiveness that it is able to rival foreign cabinets of this kind both in usefulness for the sciences [*wetenschappen*; similar to the German *Wissenschaften*] and in the way in which the preparations are displayed.[84]

As this phrase shows, Sandifort was interested in collections useful for research ('usefulness for the sciences'); in Chapter 1, we saw that he regularly used the collections in teaching. Yet to him, use in research and teaching did not preclude a presentation strategy that appealed to lay visitors as well.

Sandifort would be the last curator for whom this was the case: his successors, starting with Halbertsma, thought it impossible to combine the interests of students, researchers, and lay visitors. They considered collections that appealed to lay visitors to be 'unscientific', as becomes apparent from the inaugural lecture of Teunis Zaaijer, the Cabinet's last nineteenth-century curator. Although he became a curator in 1877, he had already been appointed as a professor in anatomy 12 years earlier. In his inaugural lecture, he fiercely criticized the most famous Dutch early modern anatomist, Frederik Ruysch. According to Zaaijer, '[Ruysch has] shown, through the layout of his collections, that he missed the true method, the right scientific genius; he made anatomy, as it were, a fashionable product for the great of the earth.'[85] Like

Halbertsma, Zaaijer suggested that one could *either* be 'scientific' and thus useful for research and teaching, *or* please the lay public (in this case, 'the great of the earth' – which could be a nod to Peter the Great, the most famous visitor to Ruysch's collection). In Zaaijer's eyes, Ruysch had chosen the latter, and this annoyed him:

> Anatomy owes Ruysch some important improvements, but we cannot get away from the conviction that through a better method, such a long and productive life, almost all of it in good health, could have resulted in more fruits for our science [of anatomy].[86]

Other nineteenth-century anatomists criticized Ruysch for similar reasons. Joseph Hyrtl, for example, stated in 1860 that the fame of Ruysch's collection was mainly due to 'curiosities' such as his *memento mori* tableaus with tiny skeletons on burial mounds made of bladder stones, and that it had little to do with Ruysch's scientific merits, 'which indeed did not rank among the most impressive ones'.[87] Zaaijer and Hyrtl were right insofar that Ruysch's collection offered entertainment for lay audiences, including noble and royal visitors, but they forgot that Ruysch also actively used his preparations in teaching and research.[88] To Ruysch and other early modern anatomists, this was a natural combination. To Zaaijer, Halbertsma, Hyrtl, and their contemporaries, it was an impossible one.

In the second half of the nineteenth century, scientists – anatomists, but also physicists, chemists, biologists, geologists, and so forth – increasingly distinguished between 'scientific' and lay audiences. They aimed to create a strict separation between the two, which helped them acquire authority in society. Andrew Cunningham has pointed to this rise of expertise in the natural and life sciences as one of the developments relating to the disappearance of the public dissection; it also relates to the closing off of anatomical collections.[89] To create and maintain their status as experts, scientists had to distinguish themselves from 'amateurs', which is how that word acquired the negative connotation it carries today. Thus, scientists labelled themselves 'scientific' and everyone else 'amateurs', and then presented the two categories as mutually exclusive. Books, exhibitions, and other works on natural knowledge aimed at laypersons were called 'popular science', where 'popular' had the negative connotation of being 'non-scientific'.[90] As a result, things such as, say, collections could no longer be 'popular' and 'scientific' at the same time. Excluding lay visitors became a means to present collections, and with them their owners, as scientific. The exclusion could be formal, by not allowing visitors to enter the collections at all.[91] But it could also be (slightly) subtler, by impressing visitors with incomprehensible preparations and thus convincing them that medical knowledge and the authority that accompanied it were best left to the professionals.[92]

In Leiden, as we have seen, lay visitors were not officially deterred; they could come and visit the collections if they wanted to. But the collection's curators, in their ambitions to be scientific, focused exclusively on students and researchers. From 1879 onwards, they were even required to do so by law. The 1879 decree on the management and use of collections in higher education stated that curators could allow visitors only 'so long as this does not cause any trouble for its [the collection's] intended use. As soon as teaching concerns or the institution's interests make it necessary, visitors should be refused.'[93] The curators no longer made any effort to help lay visitors relate to the preparations. As we have seen, lay visitors made sense of preparations by adding their own stories and knowledge. To do so, they required a point of departure to which they could tie such stories. Before 1860, points of departure had been abundant; after 1860, they disappeared. Being part of a university collection, the new Cabinet's preparations were specialized in nature. Since anatomy had lost its religious and moral aspects, the preparations were now solely intended for teaching and researching the structure of the body. As a result, people without medical knowledge struggled to understand them. They needed tales on tablets or stories told by guides in order to see more than just shelves full of medical objects – to see the son of a senile usurer, the head of a famous giant, and the skeletons of criminals. But the curators made no effort to provide such tales, and religious and moral issues no longer formed a natural part of the discipline of anatomy. Hence, late nineteenth-century visitors were confronted not with interpretable preparations, but with collections to which they could hardly relate.

We can reconstruct quite accurately what the few remaining visitors would have encountered when they entered the rooms housing the anatomical collections in the new building. The arrangement of the preparations becomes clear from a handwritten inventory that lists the preparations by cupboard. The inventory was compiled by Zaaijer, who sent it to the governors in January 1893.[94] Of course, the collections were regularly extended between 1860 and 1892, which means that not all preparations mentioned in the inventory would have been visible throughout the period. Furthermore, at two points in time, large parts of the collections were removed: in 1861 part of the Brugmans collection was moved to the natural history museum, and in 1885 many of the pathological preparations went to the new pathology laboratory. But we have no reason to assume that the arrangement of the remaining preparations was changed significantly. Except for the addition of a gallery in 1867–1868, the annual reports make no mention of extensions or changes in the collection rooms, whereas changes in other anatomy rooms are discussed in some detail.

According to the 1892 inventory, four of the Cabinet's rooms were dedicated solely to the collections: rooms 9 to 12. (Some of the other rooms, such as the preparation room and the curator's office, contained preparations as well; they were not included in the inventory.)[95] Room 9 was the most varied and contained wet and dry preparations of comparative anatomy, developmental history, and human anatomy. The room contained 10 large cabinets and 12 smaller ones, most of them with over a hundred preparations. Cabinet IV, for example, contained 252 fluid preparations on human anatomy: 80 on skeletal development; 55 of skin, nails, and hair; 41 of the senses; and 76 of the digestive system.[96] None of these were likely to have been of much interest to lay visitors. Moreover, even if visitors had been able to understand the preparations of the digestive system, one or two would have been more than enough. Visitability was certainly not aided by having 76 preparations of the same kind.

The most interpretable preparation in room 9 – and in the Cabinet as a whole – was probably the 'mice orchestra'. The orchestra was an impressive piece of handiwork by the Dutch doctor E. J. van der Mijle. Van der Mijle had collected enough mice skeletons to put together a miniature orchestra, which he then donated to the Anatomical Cabinet. In the accompanying letter, he stated his intentions:

> I hope that the gloominess associated with anatomical cabinets will disappear thanks to the musicians' tuneful tones and the truly musical touch with which they handle their instruments; and [I hope] that the visitor, nervously melancholic due to various unpleasant sensations, will see his previous cheerful mood return.[97]

If used in this way, the orchestra would have undoubtedly made the collections more visitable. But the piece was put on top of a large cabinet; not a place where it would easily catch a visitor's eye. Apparently, the preparation was not considered part of core scientific business, and the Leiden curators could hardly be bothered with countering the unease felt by visitors.

Rooms 10 and 12 were largely filled with anthropological skeletons and skulls. Until 1885, room 10 had housed the pathological preparations as well. When these had been moved to the pathological laboratory, some of the anthropological preparations from room 12 (which suffered from a lack of space) were rehoused. Room 10 also contained some 'ordinary' skeletons. According to the inventory 'the skeletons are marked A to V; on the skulls have been written the sex and, wherever possible, the age'.[98] Twenty-two skeletons, but none of them held banners warning that life was short, nor were they individualized by tales of the crimes they had committed. Instead, they were nameless, reduced to their sex and, when known, their age. To the non-medical gaze, all of them would have looked the same.

The remaining room, room 11, contained 24 cabinets (12 large, 12 small), all filled with teratological preparations. If a mother wanted to visit her malformed child, she had to come to this room. But she may not have been able to come as close to the child as she had been in the old Cabinet. We do not know the extent to which lay visitors in the new Cabinet were allowed to come close to, or even touch, the preparations, but the policy was probably more restrictive than it had been in the old Cabinet. At least, that is what we know of other anatomical collections of the time (if they admitted lay visitors at all): handling by lay visitors was increasingly discouraged, or even explicitly prohibited.[99]

By the end of the nineteenth century, most institutional anatomical collections had become hard to enter for lay visitors. Their inaccessibility contrasted sharply to other types of collections, which by this time had opened up to a wide audience. In some places, visitors without a medical background were officially banned. In others, such as Leiden, visits ceased not because visitors were explicitly sent away, but because the collections had lost their visitability. As medical practices and attitudes changed, anatomical preparations ended up in laboratories that were difficult to approach, organized in 'scientific' arrangements that made them hard to interpret without prior medical knowledge.

The afterlife of the wrinkled boy

What happened to the malformed infant of the beautiful young woman and the ugly old man, the child with whom we started this chapter? The child was probably among the preparations that were moved to the new pathology laboratory in 1885.[100] Two facts support this claim. First, the preparation currently carries a label from the pathology laboratory, which indicates that the laboratory possessed it at some point. Second, the preparation is not listed in the extensive catalogue of teratological preparations in the Anatomical Cabinet that was compiled in 1910, meaning it was no longer at the Cabinet at that time.[101] Unfortunately, we cannot look the preparation up in the pathology lab's collection catalogue, as the label has become illegible over time. The undated pathology catalogue was probably compiled in the early twentieth century. It is concise, with the preparations being described in one or two words.[102] Several of the descriptions would have fitted the monstrous child: 'monstrum' or 'foetus', for example. But it was most likely described as an 'anencephalus'. In an anencephalus, (part of) the skull is missing, and the brain is absent or deteriorated; this is the major malformation shown by the preparation. Whether the preparation was moved to the pathological laboratory or remained in the new Cabinet, it was this malformation that would have been used to characterize it – not the story of its parents.

These days, the monstrous child is housed in the Anatomical Museum of the Leiden University Medical Center. The last time I saw it, about six years ago, it was lying in a drawer in one of the museum's storage rooms, in the basement of a medical teaching building. It carried an illegible label and the museum database described it as 'anencephalus and rachischisis'. Like most of the anatomical preparations in Leiden, even the ones on display in the Anatomical Museum, the wrinkled boy would never again become as accessible as it had been before 1860. According to its website, the Anatomical Museum is intended for (future) medical students and their teachers;

its collections can also be used in research. Twice a year, in April and October, the museum opens its doors to 'other interested persons'.[103] But even on these two days, the museum is not exactly accessible: the building is difficult to approach and its collections are hard to interpret. The museum is housed in the university hospital's teaching building, a closed space located in the university's Bio Science Park. From the outside, the visitor would never guess that the building was home to a museum; and even though the museum is located close to the front entrance, it is hard to find upon entering. The museum is signposted only from the back entrance; the lack of signs outside and at the front entrance is sometimes remedied with temporary signs that are put out on days that the museum is officially open to the general public. The museum entrance is located in a dead-end corridor and the glass door has been made non-transparent. Moreover, once inside, the lay visitor will find it hard to interpret the preparations. Touch screens offer information about individual objects, but the texts, which contain many medical terms, speak to a specialist audience. The guides are medical students; their tours are hard to follow without medical knowledge.

The Leiden collections are far from being the only present-day anatomical collections that are open to lay visitors in theory, but hard to enter in practice. A remarkable example, even for medical museums' standards, is the National Museum of Health and Medicine in Washington, which holds anatomical preparations from both military and civilian sources. Unlike other US national museums, it is not located at the National Mall, but at an army base in the suburbs. Nonetheless, visitors do come to the museum. It is open to the public, just like many of the anatomical museums housed in academic hospitals, which have absorbed the nineteenth-century research and teaching laboratories to which the museums were relocated. Yet, although visitors are allowed to enter them, these museums are nowhere near as accessible as the average art museum. The limited accessibility is due mainly to their distant location, which is often (but not always) paired with limited opening hours and a presentation style directed more at medical students than at lay visitors.

The exact path travelled to end up in such locations differed for each collection. Some collections were moved to teaching laboratories in the mid-nineteenth century; others, to research laboratories in the early twentieth century. Some collections underwent abrupt changes; at others, rearrangement and relocation happened gradually. Some collections were formally closed to the lay public, only to (partially) re-open well into the twentieth century. Others were never completely closed, and a few even continued to welcome visitors actively throughout the nineteenth century. But almost all of these individual paths were shaped by a shared characteristic: the continued usefulness of anatomical preparations in medical research and teaching throughout the nineteenth century, and beyond. Hence, the medical faculties took them wherever they went – usually far away from other tourist destinations, both in distance and in style.

Had all anatomical collections lost their (medical) use in the nineteenth century, more of them might have ended up in easily accessible spaces; not as medical objects, illustrating the structure of the body, but as historical artefacts, telling us about cultures past. This happened to other types of scientific objects, such as chemical instruments or geological models – insofar as they were not thrown away before they reached public museums. Occasionally, it happened to anatomical preparations too, including in Leiden, where a few historical preparations are currently on display in Museum Boerhaave, a museum for the history of science and medicine. Yet most preparations resisted such historization. They lost their connection to the past, just as they lost their moral stories – much to the dismay, as we will see, of our next audience: the university governors.

Notes

1   *Billets*, 51–52.
2   See for example Kong et al., 'Rate'.
3   MacGregor, *Curiosity and Enlightenment*, 161–62; Rupp, 'Matters of Life and Death', 264; Zuidervaart, '"Theatrum Anatomicum" te Middelburg', 78.
4   Hildebrand [Nicolaas Beets], *Camera obscura*, 116–17.
5   *Billets*, 54.
6   Zuidervaart, 'Academische schouwplaatsen', 15–16.
7   MacGregor, *Curiosity and Enlightenment*, 161.
8   Jorink, *'Boeck der Natuere'*, 287.
9   On the book-of-nature metaphor, see Jorink, *Reading the Book of Nature*; on the use of the metaphor by early modern Leiden anatomists, see Huisman, *The Finger of God*.
10 Bakhuizen van den Brink, *De Nederlandse Belijdenisgeschriften*, 73; translation taken from Huisman, *The Finger of God*, 57–58.
11 Hooke, *Posthumous Works*, 338.
12 Minutes of the governors, 8 February 1850, file 36, Archief van Curatoren 1815–1877 (hereafter cited as AC2), Leiden University Library.
13 Halbertsma to university governors, 4 April 1850, file 113, document 81, AC2.
14 Huisman, 'Resilient Collections'; Knoeff, 'The Visitor's View'.
15 Knoeff, 'The Visitor's View'; Dicks, *Culture on Display*.
16 Dicks, *Culture on Display*, 1.
17 *Billets*, 52.
18 Knoeff, 'Touching Anatomy'.
19 'Professor Brühl's erste diesjährige Sonntagsvorlesung', 116; translation taken from Buklijas, 'Dissection', 155.
20 'Notizen', 508; translation taken from Buklijas, 'Dissection', 155.
21 Alberti, 'The Museum Affect'.
22 *Billets*, 52–53.
23 Knoeff, 'The Visitor's View'.
24 Blancken, *Catalogue*, 4, 5, 10.
25 Bennett, *Birth of the Museum*, 92–93.
26 Tibbe and Weiss, 'Druk bekeken'.
27 Bergvelt, 'Tussen geschiedenis en kunst', 345.
28 Bergvelt, 346.

29 Janse, 'Out of Curiosity', 12.
30 Engel, *Hendrik Engel's Alphabetical List*, 88, 279; Rupp, 'Matters of Life and Death', 264; Zuidervaart, '"Theatrum Anatomicum" te Middelburg', 78–109.
31 Guerrini, 'Duverney's Skeletons', 592; MacGregor, *Curiosity and Enlightenment*, 161–62.
32 MacGregor, *Curiosity and Enlightenment*, 161.
33 MacGregor, 161.
34 Cunningham, 'End of the Sacred Ritual'; Rupp, 'Het theatrum anatomicum'. Dissections could have more functions besides those of religious ritual and moral-philosophical lesson: they also boosted the status of the city and the university, and their strict regulations disciplined the audience. On status, see for example Ferrari, 'Public Anatomy Lessons and the Carnival'; on the disciplining of medical students especially, see Klestinec, *Theaters of Anatomy*. However, these functions do not concern us here, because they were not so much what made the event understandable to a wide audience, as what made it attractive to a small group of organizers. A word of caution: not all of these interpretations of public dissections are applicable throughout Europe. On this, see in particular Klestinec, who convincingly argues that the famous Padua theatres require a different interpretation than the ones usually offered by historians of medicine.
35 On public dissections as punishment, see Sawday, *The Body Emblazoned*, 54–58.
36 Alberti, *Morbid Curiosities*, 173–74.
37 Cunningham, 'End of the Sacred Ritual'.
38 Cunningham, *The Anatomist Anatomis'd*, 361–89.
39 On laboratories, see Cunningham and Williams, *Laboratory Revolution*; Jackson, 'The Laboratory'.
40 Annual reports of the Anatomical Cabinet 1850–1851 and 1852–1853, file 270, AC2.
41 Annual report of the Anatomical Cabinet 1851–1852, file 270, AC2.
42 I would like to thank Harm Beukers for suggesting this explanation.
43 See for example the Cabinet's annual reports of 1853–1854 and 1855–1856, file 270, AC2.
44 Beukers, 'Groei en ontwikkeling', 93. On the differences between Sandifort's and Halbertsma's teaching methods, in particular their (non-)use of microscopes, see Beukers, 'Beginjaren van de microscopie'.
45 Cunningham, 'Old Physiology'; see also Nyhart, *Biology Takes Form*, 67–80.
46 Otterspeer, *Wiekslag*, 119–23.
47 Otterspeer, 122.
48 Huizinga, 'De academische gebouwen', 24–26.
49 'Het laboratorium', 64.
50 Annual report of the Anatomical Cabinet 1860–1861, file 271, AC2. Unfortunately, I have not found any other sources on the opening lecture to which Halbertsma refers.
51 Leidsch Studenten Corps, *Studenten-almanak 1860*, 164.
52 Leidsch Studenten Corps, *Studenten-almanak 1861*, 226–27.
53 Annual report of the Anatomical Cabinet 1867–1868, file 272, AC2.
54 Annual report of the Anatomical Cabinet 1883–1884, file 1553, Archief van Curatoren 1878–1953 (hereafter cited as AC3), Leiden University Library.
55 Van Delft, *Heike Kamerlingh Onnes*, 178–89.
56 'Bezoekalbum van het Anatomisch Kabinet te Leiden', 1860–1876, file 1842, Bibliotheca Publica Latina Collection, Leiden University Library.
57 Visitors numbers for the RCS museum can be found in the triennial reports of the boards of curators (Reports from the Boards of Curators, 1800–1822, file 8/3/1, Papers of the Hunterian Museum and the Wellcome Museum [hereafter

cited as RCS-MUS], Royal College of Surgeons Archives, RCS-MUS, London), the minutes of the museum committee (summarized in Keith, *Abstract of Minutes*), the minutes of the Hunterian Trustees (extracts published in Negus, *Trustees*), and 'regular' visitor books (Visitor Books for the Royal College of Surgeons' Museums, 1819–1970, file 6/1, RCS-MUS).

58 'Register of illustrious and distinguished visitors', 1805–1932, file 6/2/1, RCS-MUS.
59 Nys, *De intrede van het publiek*, 74.
60 Halbertsma, *Scholars, Travellers, and Trade*, 145–47.
61 Alberti, *Morbid Curiosities*, 172–74.
62 Häner, 'Restoration Reconsidered', 254.
63 Bennett, *Birth of the Museum*.
64 Bates, '"Indecent and Demoralising Representations"'; Burmeister, 'Popular Anatomical Museums'; Claes and Deblon, 'Van panoramisch naar preventief'; Sappol, 'Morbid Curiosity'; Stephens, *Anatomy as Spectacle*.
65 Burmeister, 'Popular Anatomical Museums'; Bates, '"Indecent and Demoralising Representations"'.
66 Claes and Deblon, 'Van panoramisch naar preventief'.
67 'Gemengd nieuws', *Leidsch Dagblad*, 13 April 1885.
68 Claes and Deblon, 'Van panoramisch naar preventief'.
69 Hoffmann, 'Sleeping Beauties', 141; Onghena, 'Spektakelstukken', 46–48.
70 Halbertsma to governors, 18 January 1861, file 131, document 33, AC2.
71 Leidsch Studenten Corps, *Studenten-almanak 1860*, 164.
72 Halbertsma to governors, 15 March 1864, file 137, document 71, AC2.
73 Annual report of the Anatomical Cabinet 1871–1872, file 273, AC2.
74 Halbertsma to governors, 18 January 1861, file 131, document 33, AC2 (decision governors added to the letter); governors to Halbertsma, 30 April 1864, file 461, document 112, Archieven van Senaat en Faculteiten (hereafter cited as ASF), Leiden University Library.
75 Annual report of the Anatomical Cabinet 1851–1852, file 270, AC2.
76 Annual report of the Anatomical Cabinet 1851–1852, file 270, AC2.
77 Annual report of the Anatomical Cabinet 1853–1854, file 270, AC2.
78 Annual report of the Anatomical Cabinet 1855–1856, file 270, AC2.
79 Annual report of the Anatomical Cabinet 1854–1855, file 270, AC2. The catalogues were added to the university library, see List of acquisitions of the University Library 1854–1855, file 338, AC2.
80 Senate to governors, 1 February 1854, file 119, document 138, AC2.
81 Silliman, *Journal of Travels*, 2: 164.
82 Duchesne, *Voyage d'un iconophile*, 268.
83 Van Meerten, *Reis*, 304.
84 Annual report of the Anatomical Cabinet 1837, file 270, AC2.
85 Zaaijer, *Ontleedkundige techniek*, 19.
86 Zaaijer, 20.
87 Hyrtl, *Handbuch*, 591.
88 Knoeff, 'Touching Anatomy'.
89 Cunningham, 'End of the Sacred Ritual', 203; Cunningham, *The Anatomist Anatomis'd*, 389.
90 Topham, 'Historicizing "Popular Science"'.
91 Alberti, 'Owning and Collecting', 152; Alberti, 'The Museum Affect', 380.
92 McLeary, 'Science in a Bottle', 260–70.
93 'Reglement op het beheer en het gebruik der verzamelingen, inrigtingen en hulpmiddelen voor het onderwijs aan de Universiteiten des Rijks', 31 December 1879, article 8.

94 Zaaijer, 'Inventaris der verzameling in het Anatomisch Kabinet van de Rijks Universiteit te Leiden', 1892, archives Anatomisch Museum (no inventory number), Leiden University Medical Center; a copy of this inventory (without some of remarks written in the margins) can be found in the archives of the university governors: file 1772, AC3.
95 A separate catalogue of the wet preparations in the preparation room exists, but it is not dated. 'Katalogus spiritus-prepar. kast prepareerkamer', archives Anatomisch Museum (no inventory number), Leiden University Medical Center.
96 Zaaijer, 'Inventaris der verzameling in het Anatomisch Kabinet van de Rijks Universiteit te Leiden', 1892, p. 3, archives Anatomisch Museum (no inventory number), Leiden University Medical Center.
97 Van der Mijle to Leiden professors, 1870, file 461, ASF.
98 Zaaijer, 'Inventaris der verzameling in het Anatomisch Kabinet van de Rijks Universiteit te Leiden', 1892, p. 34, archives Anatomisch Museum (no inventory number), Leiden University Medical Center.
99 Alberti, *Morbid Curiosities*, 181.
100 Annual report of the Anatomical Cabinet 1884–1885, file 1553, AC3.
101 Van der Guyten, 'Catalogus van het Anatomisch Kabinet te Leiden', 1 October 1910, archives Anatomisch Museum (no inventory number), Leiden University Medical Center.
102 'Notulenboek Pathologie', n.d. [early twentieth century], archives Anatomisch Museum (no inventory number), Leiden University Medical Center.
103 www.lumc.nl/onderwijs/faciliteiten/anatomisch-museum/, last accessed 17 November 2017.

Bibliography

Manuscript sources

Leiden University Library, Special Collections: Archief van Curatoren 1815–77; Archief van Curatoren 1878–1953; Archieven van Senaat en Faculteiten; Bibliotheca Publica Latina Collection.
Leiden University Medical Center: Archives Anatomisch Museum.
London, Royal College of Surgeons Archives: RCS-MUS, Papers of the Hunterian Museum and the Wellcome Museum.

Printed sources

Alberti, Samuel J. M. M. 'Owning and Collecting Natural Objects in Nineteenth-Century Britain'. In *From Private to Public: Natural Collections and Museums*, edited by Marco Beretta, 141–54. Sagamore Beach: Science History Publications, 2005.
Alberti, Samuel J. M. M. 'The Museum Affect: Visiting Collections of Anatomy and Natural History'. In *Science in the Marketplace: Nineteenth-Century Sites and Experiences*, edited by Aileen Fyfe and Bernard Lightman, 371–403. Chicago: University of Chicago Press, 2007.
Alberti, Samuel J. M. M. *Morbid Curiosities: Medical Museums in Nineteenth-Century Britain*. Oxford: Oxford University Press, 2011.
Bakhuizen van den Brink, J. N., ed. *De Nederlandse Belijdenisgeschriften*. 2nd edition. Amsterdam: Ton Bolland, 1976.

Bates, A. W. '"Indecent and Demoralising Representations": Public Anatomy Museums in Mid-Victorian England'. *Medical History* 52(2008): 1–22. https://doi.org/10.1017/S0025727300002039.

Bennett, Tony. *The Birth of the Museum: History, Theory, Politics.* London: Routledge, 1995.

Bergvelt, Ellinoor. 'Tussen geschiedenis en kunst: Nederlandse nationale kunstmusea in de negentiende eeuw'. In *Kabinetten, galerijen en musea: Het verzamelen en presenteren van naturalia en kunst van 1500 tot heden,* edited by Ellinoor Bergvelt, Deborah J. Meijers, and Mieke Rijnders, 343–72. Zwolle: Waanders, 2005.

Beukers, Harm. 'De beginjaren van de microscopie aan de geneeskundige faculteiten te Utrecht en Leiden'. *Tijdschrift voor de geschiedenis der geneeskunde, natuurwetenschappen, wiskunde en techniek* 6(1983): 65–81.

Beukers, Harm. 'Groei en ontwikkeling: De Leidse faculteit der geneeskunde in het derde kwart der negentiende eeuw'. In *Een universiteit herleeft,* edited by Willem Otterspeer, 76–102. Leiden: Brill, 1984.

Billets in the Low Countries, 1814–1817: In a Series of Letters. London: Stockdale, 1818.

Blancken, Gerard. *A Catalogue of All the Cheifest Rarities in the Publick Theater and Anatomie-Hall, of the University of Leyden.* Leiden: Hubert van der Boxe, 1697.

Buklijas, Tatjana. 'Dissection, Discipline and Urban Transformation: Anatomy at the University of Vienna 1845–1915'. PhD diss., University of Cambridge, 2005.

Burmeister, Maritha Rene. 'Popular Anatomical Museums in Nineteenth-Century England'. PhD diss., Rutgers University, 2000.

Claes, Tinne, and Veronique Deblon. 'Van panoramisch naar preventief: Populariserende anatomische musea in de Lage Landen (1850–1880)'. *Negentiende eeuw* 39 (2015): 287–306.

Cunningham, Andrew. 'The End of the Sacred Ritual of Anatomy'. *Canadian Journal of Medical History* 18(2001): 187–204.

Cunningham, Andrew. 'The Pen and the Sword: Recovering the Disciplinary Identity of Physiology and Anatomy before 1800. I: Old Physiology – the Pen'. *Studies in History and Philosophy of Biological and Biomedical Sciences* 33(2002): 631–65. https://doi.org/10.1016/S1369-8486(02)00023-7.

Cunningham, Andrew. *The Anatomist Anatomis'd: An Experimental Discipline in Enlightenment Europe.* The History of Medicine in Context. Farnham: Ashgate, 2010.

Cunningham, Andrew, and Perry Williams, eds. *The Laboratory Revolution in Medicine.* Cambridge: Cambridge University Press, 1992.

Delft, Dirk van. *Heike Kamerlingh Onnes, een biografie: De man van het absolute nulpunt.* Amsterdam: Bert Bakker, 2005.

Dicks, Bella. *Culture on Display: The Production of Contemporary Visitability.* Maidenhead: Open University Press, 2003.

Duchesne, Jean. *Voyage d'un iconophile: Revue des principaux cabinets d'estampes, bibliothèques et musées d'Allemagne, de Hollande et d'Angleterre.* Paris: Heideloff et Campé, 1834.

Engel, Hendrik. *Hendrik Engel's Alphabetical List of Dutch Zoological Cabinets and Menageries.* 2nd ed. Prepared by Pieter Smit, with the assistance of A. P. M. Sanders and J. P. F. van der Veer. Amsterdam: Rodopi, 1986.

Ferrari, Giovanna. 'Public Anatomy Lessons and the Carnival: The Anatomy Theatre of Bologna'. *Past & Present* 117(1987): 50–106.

Guerrini, Anita. 'Duverney's Skeletons'. *Isis* 94(2003): 577–603. https://doi.org/10.1086/386383.

Halbertsma, Ruurd. *Scholars, Travellers, and Trade: The Pioneer Years of the National Museum of Antiquities in Leiden, 1818–1840.* London: Routledge, 2003.

Häner, Flavio. 'Restoration Reconsidered: The Case of Skull Number 1-1-2/27 at the Anatomy Museum of the University of Basel'. In *The Fate of Anatomical Collections*, edited by Rina Knoeff and Robert Zwijnenberg, 247–62. The History of Medicine in Context. Farnham: Ashgate, 2015.

'Het physisch, chemisch, anatomisch en physiologisch laboratorium te Leyden'. *Nederlandsch Magazijn*, no. 8(1859): 64.

Hildebrand [Nicolaas Beets]. *Camera obscura.* 3rd ed. Haarlem: Bohn, 1851.

Hoffmann, Kathryn A. 'Sleeping Beauties in the Fairground: The Spitzner, Pedley and Chemisé Exhibits'. *Early Popular Visual Culture* 4(2006): 139–59. https://doi.org/10.1080/17460650600793557.

Hooke, Robert. *The Posthumous Works of Robert Hooke, M.D. S.R.S. Geom. Prof. Gresh. &c.: Containing His Cutlerian Lectures, and Other Discourses, Read at the Meetings of the Illustrious Royal Society.* London: Smith and Walford, 1705.

Huisman, Tim. *The Finger of God: Anatomical Practice in 17th-Century Leiden.* Leiden: Primavera Pers, 2009.

Huisman, Tim. 'Resilient Collections: The Long Life of Leiden's Earliest Anatomical Collections'. In *The Fate of Anatomical Collections*, edited by Rina Knoeff and Robert Zwijnenberg, 73–91. The History of Medicine in Context. Farnham: Ashgate, 2015.

Huizinga, Johan. 'De academische gebouwen'. In *Pallas Leidensis MCMXXV*, 19–36. Leiden: Van Doesburgh, 1925.

Hyrtl, Joseph. *Handbuch der praktischen Zergliederungskunst als Anleitung zu den Sectionsübungen und zur ausarbeitung anatomischer Präparate.* Vienna: Braumüller, 1860.

Jackson, Catherine M. 'The Laboratory'. In *A Companion to the History of Science*, edited by Bernard Lightman, 296–309. Wiley Blackwell Companions to World History. Chichester: Wiley Blackwell, 2016.

Janse, Geert-Jan. 'Out of Curiosity and for Instruction'. In *Teylers Museum: A Journey in Time*, edited by Marjan Scharloo, 11–29. Haarlem: Teylers Museum, 2010.

Jorink, Eric. *Het 'Boeck der Natuere': Nederlandse geleerden en de wonderen van Gods schepping 1575–1715.* Leiden: Primavera Pers, 2006.

Jorink, Eric. *Reading the Book of Nature in the Dutch Golden Age, 1575–1715.* Leiden: Brill, 2010.

Keith, Arthur. *Abstract of Minutes of the Museum Committee, Royal College of Surgeons of England from 1800–1907.* London, 1908.

Klestinec, Cynthia. *Theaters of Anatomy: Students, Teachers, and Traditions of Dissection in Renaissance Venice.* Baltimore: Johns Hopkins University Press, 2011.

Knoeff, Rina. 'The Visitor's View: Early Modern Tourism and the Polyvalence of Anatomical Exhibits'. In *Centres and Cycles of Accumulation in and around the Netherlands*, edited by Lissa Roberts, 155–76. Berlin: Lit Verlag, 2011.

Knoeff, Rina. 'Touching Anatomy: On the Handling of Preparations in the Anatomical Cabinets of Frederik Ruysch (1638–1731)'. *Studies in History and Philosophy of Biological and Biomedical Sciences* 49(2015): 32–44. https://doi.org/10.1016/j.shpsc.2014.11.002.

Kong, Augustine, Michael L. Frigge, Gisli Masson, Soren Besenbacher, Patrick Sulem, Gisli Magnusson, Sigurjon A. Gudjonsson, et al. 'Rate of de Novo Mutations and the Importance of Father's Age to Disease Risk'. *Nature* 488 (2012): 471–75. https://doi.org/10.1038/nature11396.

Leidsch Studenten Corps. *Leidsche studenten-almanak voor 1860.* Leiden: Engels, [1859].

Leidsch Studenten Corps. *Leidsche studenten-almanak voor 1861.* Leiden: Engels, [1860].

MacGregor, Arthur. *Curiosity and Enlightenment: Collectors and Collections from the Sixteenth to the Nineteenth Century.* New Haven: Yale University Press, 2007.

McLeary, Erin Hunter. 'Science in a Bottle: The Medical Museum in North America, 1860–1940'. PhD diss., University of Pennsylvania, 2001.

Meerten, A. B. van. *Reis door het Koningrijk der Nederlanden en het Groothertog-dom Luxemburg, voor jonge lieden.* Vol. 5. Amsterdam: Schalekamp en Van de Grampel, 1829.

Negus, Victor. *History of the Trustees of the Hunterian Collection.* Edinburgh: Livingstone, 1966.

'Notizen'. *Wiener Medizinische Wochenschrift* 15, no. 29(1865): 507–8.

Nyhart, Lynn K. *Biology Takes Form: Animal Morphology and the German Universities, 1800–1900.* Chicago: University of Chicago Press, 1995.

Nys, Liesbet. *De intrede van het publiek: Museumbezoek in België 1830–1914.* Leuven: Universitaire Pers Leuven, 2012.

Onghena, Sofie. 'Spektakelstukken: De mise-en-scène van de wetenschap in de Belgische stad, 1860–1914'. In *Tussen beleving en verbeelding: De stad in de negentiende-eeuwse literatuur*, edited by Inge Bertels, Jan Hein Furnée, Tom Sintobin, Hans Vandevoorde, and Rob van de Schoor, 43–69. Leuven: Universitaire Pers Leuven, 2013.

Otterspeer, Willem. *De wiekslag van hun geest: De Leidse universiteit in de negentiende eeuw.* Hollandse historische reeks 18. The Hague: Stichting Hollandse Historische Reeks, 1992.

'Professor Brühl's erste diesjährige Sonntagsvorlesung, abgehalten im Sammlungssaale des zootomischen Universitäts-Institutes. (Nach einer revidirten stenographischen Aufzeichnung.)'. *Wiener Medizinische Wochenschrift* 16, no. 7(1866): 115–17.

Rupp, Jan C. C. 'Matters of Life and Death: The Social and Cultural Conditions of the Rise of Anatomical Theatres, with Special Reference to Seventeenth-Century Holland'. *History of Science* 28(1990): 263–87.

Rupp, Jan C. C. 'Het theatrum anatomicum: Publiekscommunicatief fossiel of "archetype"'. *Gewina* 25(2002): 191–209.

Sappol, Michael. '"Morbid Curiosity": The Decline and Fall of the Popular Anatomical Museum'. *Common-Place* 4, no. 2(2004). www.common-place.org/vol-04/no-02/sappol/.

Sawday, Jonathan. *The Body Emblazoned: Dissection and the Human Body in Renaissance Culture.* London: Routledge, 1996. First published 1995.

Silliman, Benjamin. *A Journal of Travels in England, Holland, and Scotland, and of Two Passages over the Atlantic, in the Years 1805 and 1806.* 2nd ed. Vol. 2. Boston: T. B. Wait, 1812.

Stephens, Elizabeth. *Anatomy as Spectacle: Public Exhibitions of the Body from 1700 to the Present.* Liverpool: Liverpool University Press, 2011.

Tibbe, Lieske, and Martin Weiss, eds. 'Druk bekeken: Collecties en hun publiek in de negentiende eeuw'. Special issue, *De negentiende eeuw* 34, no. 3(2010).

Topham, Jonathan R. 'Introduction to Focus: Historicizing "Popular Science"'. *Isis* 100(2009): 310–18. https://doi.org/10.1086/599551.

Zaaijer, Teunis. *Het gewigt eener doelmatige ontleedkundige techniek*. Leiden: Hazenberg, 1866.

Zuidervaart, Huib J. 'Academische schouwplaatsen en hun collecties: Het begin van de Nederlandse universitaire verzamelingen'. In *Universitaire collecties in Nederland: Nieuw licht op academisch erfgoed*, edited by Tiny Monquil-Broersen, 11–20. Zwolle: Waanders, 2007.

Zuidervaart, Huib J. 'Het in 1658 opgerichte "Theatrum Anatomicum" te Middelburg: Een medisch-wetenschappelijk en cultureel convergentiepunt in een vroege stedelijke context'. *Archief: Mededelingen van het Koninklijk Zeeuwsch Genootschap der Wetenschappen*, 2009, 73–140.

4 Adieu Albinus

How the university governors lost their status symbol

Collections make excellent status symbols for administrators looking to boost their institution's standing. In 1799, the members of the London Company of Surgeons accepted the collection of the deceased surgeon-anatomist John Hunter, not because they required it for research or teaching, but because they thought it might help their reputation. And help it did: only a few months later, the king granted the institution the 'scientific' title of Royal College. The new Royal College of Surgeons continued to derive status from its collections.[1] In the nineteenth century, its museum dazzled contemporaries: rooms filled with tens of thousands of preparations, arranged according to the newest scientific insights, and partly created by the famous John Hunter. The College's collections impressed with their size, their scientific value, and their history; many institutions had to make do with collections that were remarkable in one, or at best two, of these three areas. In nineteenth-century Leiden, the status granted by the anatomical collections hinged on their history. Unfortunately, ongoing reinterpretation of the preparations in research and teaching disconnected the objects from the famous anatomists who had made them. The preparations thereby lost their past – and the university administrators, their status symbol. This chapter tells how this happened.

To do so, it will first examine how, in the early nineteenth century, the preparations' past became crucial for their use as a status symbol. Since this is directly related to national politics (in particular, to the 1815 Royal Decree on Higher Education), the chapter starts with an overview of Dutch political developments in the early nineteenth century and their impact on Leiden University. I will show how the 1815 decree damaged the reputation of Leiden's anatomical collections and threatened the university's status. As we shall see, the university governors responded to this threat not only by extending the collections, but also by using the collections' past to position themselves above the other Dutch universities. We will then see how this historical value disappeared after the move to the laboratory in 1860. As a result, the governors lost their status symbol – and, just like the lay visitors in the previous chapter, by the end of the nineteenth century they had ceased to use the anatomical collections.

The Netherlands and Leiden University in the early nineteenth century

In December 1794, the French general Jean-Charles Pichegru had a stroke of luck. He intended to invade the Dutch Republic, but until then, his troops had been halted by that quintessential Dutch defence: water – in this case, the rivers Maas and Waal, hard-to-cross natural barriers. Pichegru was wondering what to do next when a sharp frost descended – and stayed. Two days after a cold Christmas, the French troops marched over thick ice into the Republic, marking the start of what would become known as *de Franse tijd*, the French period. This period, which ended in 1813, was characterized by changes of government. In early 1795, the Batavian Republic was established as a sister republic of France. In 1806, the French turned the new Republic into the Kingdom of Holland and Napoleon Bonaparte appointed his younger brother Louis as king. Napoleon had a habit of putting family members in charge of vassal states; this guaranteed his direct influence and maintained a gloss of independence and legitimacy.[2] This time, though, his plan backfired. Only four years after Louis's crowning, Napoleon felt forced to invade the kingdom to reclaim his power. His brother had systematically put Dutch interests above those of France; he even used the Dutch version of his name, Lodewijk. In particular, Louis had refused to acknowledge Napoleon's demands for money and soldiers. As a result, Napoleon annexed the kingdom into the French Empire. After the defeat of Napoleon, the borders of Europe were redrawn by the European powers. In Northwest Europe, the Kingdom of the Netherlands (which included what we now know as the Benelux countries) was created to form a buffer against France. The son of the last stadtholder became the ruler of this new country and, in 1815, the first Dutch king: William I.

William's kingdom had been designed on the drawing board. Some of its component regions had cooperated before, but rather loosely. In the Dutch Republic, which roughly overlapped with the new kingdom's northern provinces, most business had been done locally. The new country was diverse; it was characterized by its differences, not by its similarities.[3] First of all, the same rivers that had slowed down Pichegru divided the country into two regions, the North and the South. There was hostility between the two; the South – justifiably – felt looked down upon by the North. A wide gap also existed between the cities and the countryside. Furthermore, the population was divided in religious terms. Protestants and Catholics opposed one another (or rather, Protestants stood above Catholics), and within Protestantism, bitter conflicts would result in dozens of denominations. Last, wealth and income inequality and class consciousness had grown stronger after the economic decline in the second half of the eighteenth century.

Despite all these differences, William was determined to turn his new country into a unitary state.[4] To do so, he pursued a policy that was both centralizing and unifying. He centralized government to such an extent that he made many decisions by himself, including detailed ones; he dealt with everything.[5] The parliament had little control over the king and thus over the country. His attempts to unify the kingdom included making Dutch the national language (at the expense of French, the main language in some of the southern provinces). His aim to unify the new state also shaped his economic and educational policies, including the decree on higher education.[6] William, who liked to think of himself as the *landsvader*, 'the father of the nation', aimed to love all of his children equally.

But some of his children considered themselves more equal than others – among them Leiden University's governors. They were neither used to, nor fond of, being unified. After its foundation in 1575, their university had quickly gained an international reputation. Its anatomical theatre, botanical garden, and library attracted students and scholars from all over Europe, as did professors such as Pieter Pauw (1564–1617), Carolus Clusius (1526–1609), and Joseph Scaliger (1540–1609). Leiden University was a centre of excellence in Europe, and it remained so until the late eighteenth century. In 1765 the *Encylcopédie* even declared it the first (i.e., the best) university in Europe: 'The university of Leiden is the first of Europe. It seems that all famous men in the republic of letters went there, allowing it to flourish from its establishment until our day.'[7]

By then, the medical faculty was responsible for a large part of the university's fame, with celebrated professors such as Herman Boerhaave and Bernhard Siegfried Albinus, and with the well-known Leiden anatomical collections. The anatomical theatre had contained a small collection of anatomical objects since the late sixteenth century. In the eighteenth century, important additions were made to the collection. Early in the century, Leiden anatomy professor Johannes Rau (1668–1719) bequeathed his preparations to the university. The governors welcomed the gift and asked the newly appointed Bernhard Siegfried Albinus to catalogue the collection. In addition to managing the university's collections, Albinus built a large private collection for his research and teaching. In 1771, a year after Albinus's death, the university acquired this collection as well. The governors asked two medical professors, Eduard Sandifort and Frederik Bernhard Albinus (Bernhard Siegfried's brother), to write a report on the Albinus collection. The professors did as they were asked, and took their chance to ask the governors for some additional money. They intended to reorganize the older anatomical collections exhibited in the Anatomical Theatre, which, in their eyes, had been neglected. They wrote:

> The costs [of reorganizing the anatomical collections displayed in the Anatomical Theatre] are slight compared to the honour this university would gain from it, because the university would be able to pride itself not only on an excellent library, an outstanding [botanical] garden, [and] a splendid Cabinet of Natural Curiosities, but also on an Anatomical Theatre adorned with the cabinets of two famous professors [Rau and B. S. Albinus] and many other exquisite things, which would make it stand out above all others.[8]

The professors were hardly exaggerating when they claimed that the reorganized anatomical theatre, supplemented with the collections of Rau and Albinus, would be better than 'all others', certainly if by 'all others' they meant the other *Dutch* anatomical theatres. The four other Dutch universities – Groningen, Utrecht, Harderwijk, and Franeker – could in no way compete with Leiden. The student numbers confirmed this: Leiden had many more students than the other universities, although the differences diminished somewhat towards the end of the eighteenth century. Furthermore, Leiden attracted more international students than the other universities.[9] At the end of the eighteenth century, Leiden found itself in a comfortable position. But things were about to change.

The nineteenth century brought several problems for Leiden and its anatomical collections. The first arrived on a Monday morning, on 12 January 1807.[10] A powder ship berthed in Leiden's main canal, the Rapenburg, where it would remain until that afternoon. Around four o'clock, the crew started making dinner. It seems that they did not pay enough attention to the fire while cooking, because at a quarter past four exactly, the ship, which was carrying 18,500 kilograms of gunpowder, exploded. Over 200 buildings were ruined and approximately 150 people died, including two university professors.[11] All of the university's main buildings and many of the professors' houses were located on the Rapenburg; several of them were damaged or destroyed. As for the anatomical collections, the collection built by Wouter van Doeveren suffered the most damage. Leiden University had a lot of repair-work to do.

The university was generously assisted by King Louis Napoleon. His behaviour after the gunpowder disaster became a standard example of how he was much more concerned with his citizens than his brother, Napoleon Bonaparte, wanted him to be.[12] Only a few hours after the disaster, Louis arrived in Leiden, where he stayed all night to help and offered a reward for each living person extracted from the ruins. He also received a delegation of university administrators to find out what the university needed. The administrators had their priorities straight: the first thing that they asked for was not money, building materials, or replacements for the lost collections, but a new title – a clear indication of where the university stood on status.[13] The Leiden governors had long been convinced that they deserved a special title and had attempted to acquire one before: in 1800, they had asked to become the 'National Batavian University'. The government official responsible had summarized their argument as:

> the height [being the National Batavian University] ... for which it was originally meant and to which it became entitled at the time, both because of the renown that it had acquired throughout the learned world and because of the most precious collections brought together there.[14]

The governors felt they deserved a special title because of their fame and their precious collections, which included the anatomical collections. In their eyes, the collections were directly related to their status. The 1800 attempt failed, but after the gunpowder incident, Louis could not refuse the request.[15] Leiden received the epithet *Universitas Regiae Hollandiae*. The governors were delighted, as the minutes of their meeting on 4 February 1807 reveal:

> [The governors have been told that] the University [*Hoogeschool*] of Leiden will take the name Royal University [*Universiteit*] of Holland; and .... that the necessary steps will be taken to add the utmost lustre and the greatest fame to it ..., [and the governors are] imbued with an awareness of the enormous value of the boon that His Royal Majesty has given to the university, which now becomes superior to all other academies of the Kingdom and will for aye be able to flourish and shine throughout the learned world, in all the lustre for which it was originally established, and which it has deserved and maintained throughout its history.[16]

The university had lost people and buildings, but the new title, and the status that came with it, added a silver lining to the first cloud in the nineteenth-century sky.

Leiden's position was further enhanced a few years later. After Napoleon had annexed the Kingdom of Holland, he issued an Imperial Decree (1811) to restructure Dutch higher education. His centralized and hierarchical educational system had no need of all five of the universities that co-existed in the Dutch Republic, and Napoleon closed two of them: Harderwijk and Franeker. Leiden and Groningen remained fully functional, but were merged into the Université Impériale. Utrecht, Leiden's main rival, was downgraded to an *école sécundaire* (literally: secondary school), its entire staff becoming subordinate to Leiden University's Senate. The Utrecht rector was outraged at being demoted to being a 'servant of the Leiden rector'.[17] Many students left Utrecht, because the new *école sécundaire* was not allowed to award doctoral degrees. The number of students dropped from almost 200 just before the downgrading to 140 a year after.[18] Most of the remaining students were theologians, because they did not need the doctoral decree. There were only 12 students left in the medical faculty in 1813; in that same year, Leiden had 81 medical students.[19]

Thus, Leiden enjoyed a clear advantage over its chief competitor, Utrecht – but it was an advantage that it was about to lose. In 1815, a new problem arose: the Royal Decree on Higher Education.

The Royal Decree on Higher Education (1815)

In line with William I's policy, the 1815 Royal Decree on Higher Education (RDHE) both centralized and uniformized higher education. Centralization happened through shifting power from the local level (the university governors) to the national level (the Ministry of the Interior and thus the king).[20] This shift had begun in the Batavian Republic; William extended the national structures created by the French.[21] The RDHE replaced the Imperial Decree of 1811, which had been issued by Napoleon and was based on a report by Jean-François Noël and Georges Cuvier.[22] We met the latter in Chapter 2 as a zoological comparative anatomist; he was also a political advisor. The report that Cuvier and Noël had written was the third report on Dutch higher education in five years. The two earlier reports appeared in 1807 (produced by a committee led by Johan Meerman) and 1809 (under a committee led by Jean Henri van Swinden).[23] A fourth committee was established to prepare the 1815 decree, chaired by Frans Adam van der Duyn van Maasdam. This committee proposed to reverse several French measures; in particular, it wanted to return power to the local university governors.[24] But the king rejected this part of the proposal, and the centralized organizational structure of the final decree resembled the one introduced by the French four years earlier.

The RDHE made higher education more uniform in two ways. First, all universities were considered equal. The decree made Utrecht a university again, whereas Franeker and Harderwijk became *athenea* (higher education institutes ranking below the universities), which left the Netherlands with three universities: Leiden, Utrecht, and Groningen.[25] Leiden was given more professors than the other two and these professors earned higher salaries. No distinction was made in rank, however, and Leiden lost its official title. Historians, especially those writing the history of Leiden University, have sometimes claimed otherwise.[26] They quote from the draft version of the decree, which indeed declared Leiden to be the 'prime university' of the Netherlands.[27]

'Prime university' replaced *Universitas Regiae Hollandiae*, the title that Leiden had received after the gunpowder disaster. The governors had naturally hoped for the continuation of their official status as the premier university of the Netherlands, and the prospect of this had seemed likely. One of the most influential members of the RDHE's preparatory committee, Jan Melchior Kemper, was a Leiden professor; the committee's chair, Van der Duyn van Maasdam, was a Leiden university governor between 1813 and 1848.[28] Kemper and Van der Duyn van Maasdam were probably responsible for getting the university's premier status into the draft version of the decree. The other universities successfully opposed this decision, however, upon which the king removed the term from the final decree.[29] This must have been painful for Leiden, because the governors had assumed they had a special relationship with King William I. After all, his ancestor William of Orange had founded the university in 1575. But the king cared little for special relationships and prime universities – he wanted uniformity instead.

The second means of creating this uniformity was the detailed rules all universities had to follow. All universities had to teach the same courses. Furthermore, all university collections became similar, because they had to comply with the standards set out in the RDHE. One of the decree's sections was devoted to 'material assistance for academic teaching'.[30] It prescribed which material assistance should be present – including several collections, a library, a chemistry laboratory, and an observatory. Furthermore, it regulated the management of these objects and institutions. For medical teaching, it prescribed an academic hospital and collections of surgical and obstetrical instruments, medical books, and anatomical preparations.[31] Article 177 specified the contents of the collections with anatomical preparations:

> At all universities there will be cabinets of anatomical, physiological, and pathological preparations and objects, for the assistance and advancement of the teaching of anatomy, medicine, surgery, and obstetrics; to these cabinets will also be added such preparations of *anatome comparata*, as can serve to elucidate the knowledge of the human body.[32]

This requirement and the policy that William I based on it threatened the Leiden anatomical collections, because it made them both less adequate and less unique.

Although the university possessed a rich anatomical collection, it did not comply with the decree's demands. In their first annual report following the decree, the governors admitted that their collections were incomplete:

> The cabinets for the advancement of the teaching of anatomy, medicine, and obstetrics are to varying degrees equipped with anatomical, physiological, and pathological preparations and objects – although not to the extent required; and the name of Albinus, whose cabinet belongs to the possessions of the university, may lead one to expect much; we would, however, be failing to honour the truth if we were to assure your Excellency [the Minister of Education] that Leiden reaches the standards of science in this respect, and that there are no needs, even more so because the *Anatome Comparate*, valued properly by the Royal Decree, leaves much, if not everything, to be desired.[33]

According to the governors, the collection was especially lacking in comparative anatomy preparations; something that is confirmed by the first two volumes of the collection catalogue *Museum Anatomicum Academiae Lugduno-Batavae*, published in 1793.[34] Since the university did not acquire many new preparations between 1793 and 1815, the volumes give an adequate overview of the preparations at the time the decree was issued. The volumes list around 2,500 preparations.[35] Most of them were general-anatomical, some pathological, very few comparative-anatomical. The Albinus collection, for example, contained 752 preparations, of which only 66 were listed as animal preparations.[36] The collection of Wouter van Doeveren consisted of 441 preparations, only 15 of which were animal preparations.[37] Even if we assume that all of these animal preparations were comparative-anatomical, their number was small. Moreover, some of the animal preparations should probably be listed as natural-historical, because they showed the outside of the animal (such as stuffed animals, or whole animals in fluid) and not, as comparative anatomy preparations did, its internal structures.

Eighteenth-century anatomists such as Albinus and Van Doeveren had no need to include comparative-anatomical preparations in their collections: comparative anatomy was not introduced in Dutch university teaching until the end of the eighteenth century. Sebald Justinus Brugmans was the first Leiden professor to teach comparative anatomy. As we saw in Chapter 2, he built an impressive anatomical collection with at least 2,000 comparative-anatomical preparations, which he used in his teaching. Brugmans was appointed professor in Leiden in 1785; in 1815, his collection had more or less reached its full size. Thus, when the RDHE was issued, a large comparative anatomy collection was available for teaching medical students. However, this collection was not owned by the university, but by an individual professor, Brugmans. For this reason, the governors could not claim that the collection met the demands of the decree.

In the early modern period, most collections (anatomical and otherwise) were privately owned; Leiden's large institutional collections were an exception. But what had previously been exceptional became standard in the nineteenth century, when collection ownership shifted from private to institutional hands.[38] The Dutch government stimulated this shift, as it seems to have discouraged professors and curators from building private collections. An early nineteenth-century educational report explicitly stated that the 'usefulness' of professors' private collections would become 'more general' if these collections were to become university property. This report advised the king (Louis Napoleon) to buy these collections and donate them to the universities, which, as we will see, was exactly what William I would do.[39] Much later, in 1859, the government would explicitly prohibit the directors and staff of the National Museum for Natural History from building their own collections.[40] Such explicit rules were probably intended to avoid conflicts of interest: if museum staff had their own collections, they might be tempted to use resources that belonged to the museum, such as incoming dead animals and preparation fluid. Furthermore, the government preferred institutional collections because they guaranteed continuity: collections no longer disappeared when a professor moved to a different university, or died. Institutional collections also gave the government more control over what was in the collections, and institutional collections could be made equally accessible to *all* professors. Consider the Brugmans collection as an example: when it was still private, it was located in Brugmans's home, and it was entirely up to Brugmans whether he let other professors use his preparations. As soon as it became institutional, its use was regulated by the RDHE, which clearly stated that all professors were allowed to borrow preparations from the collections. There was still only one curator, but he had to follow the rules; and if he didn't, his colleagues could complain to the governors, who had the power to overrule him.

The Brugmans collection's presence in Leiden did not suffice to fulfil the decree's demands, because the university did not own it. Hence, after the decree was issued, Leiden's collections suddenly looked (and were) deficient. Since the collections had been a major status symbol, this was painful. But Leiden still had one major advantage: the university did own a collection, which was more than the other universities could claim. Neither Groningen nor Utrecht possessed any anatomical preparations in 1815. However, Leiden's advantage would soon disappear. Only a year after the decree was issued, Utrecht acquired the Bleuland collection, a high-quality collection with many comparative-anatomical preparations.[41]

Utrecht received this collection from King William. It was by no means the last anatomical collection he donated to a university. Between 1815 and 1835, he bought at least seven collections and divided them between Leiden, Utrecht, Groningen, and Ghent.[42] (Ghent was one of the southern universities that were part of the Netherlands until 1830, when Belgium seceded.) These donations reflected William's unifying policy – and his habit of occupying himself with detailed decisions.

William's donations made Leiden's collections less unique. What was worse, the university's main rival, Utrecht, now owned something that Leiden lacked: a comparative anatomy collection. The governors felt outdone. Their collections seemed threatened with both inadequacy and a lack of uniqueness. How would they respond?

The first strategy: claim to comply with the standards

Like all Dutch universities, Leiden had five governors who administered the university. They were appointed by the king. Each university also had a senate, an assembly of professors, but their role was mainly advisory; ultimately, the governors made the decisions.[43] The governors' responsibilities included implementing the educational laws, managing the finances, and caring for buildings and collections. In the first half of the nineteenth century, the most influential governors in Leiden were chairman Frans Adam van der Duyn van Maasdam (governor from 1815 to 1848), Hendrik Collot d'Escury (governor from 1815 to 1844), and Frans Godert Lynden van Hemmen (governor from 1823 to 1845).[44] Both Van der Duyn van Maasdam and Lynden van Hemmen were members of the committee that drafted the RDHE.

The governors had a clear idea of what their main task should be. In 1822, they wrote to the minister: 'To the obligations which have been imposed on us belongs also, in particular, the promotion of everything which could serve to maintain the university's fame.'[45] Maintaining the university's fame was indeed one of the tasks assigned to the governors in the RDHE.[46] But it was the last task in a list of seven, which does not particularly justify singling it out as the most important one. And yet, the Leiden governors claimed time and again that maintaining, or boosting, the fame of the university was their main concern.[47]

The anatomical collections were a means to this end, and to use them as such, Leiden needed to convince others they were superior. To communicate this message, it was neither necessary nor sufficient to own the best collection; but it would make the job easier, which is why the governors set out to complete their collections. Several preparations were added every year, but the two most important extensions were the Brugmans and the Bonn collections. The Brugmans collection was acquired in 1819. Half of the approximately 4,000 preparations concerned comparative anatomy, the other half, pathology and natural history. Three years after it bought the Brugmans collection, the university acquired the preparations of the Amsterdam anatomist Andreas Bonn (1738–1817).[48] Bonn's collection was bought by the king and then donated to Leiden University, on the condition that preparations that duplicated those already present in the Leiden collections would be sent on to other universities.[49] Gerard Sandifort assessed the preparations.[50] He selected 737 preparations for the Leiden collections; the remaining ones were sent to the University of Ghent. Most of the Bonn preparations added to the Leiden collections involved general anatomy or pathology; some concerned comparative anatomy. Sandifort was particularly pleased with the pathology additions, specifically the monstrosities and the pathological bone preparations.[51]

After the acquisition of the Bonn collection, the university collections fully complied with the standards set in the RDHE. The Brugmans collection remedied the lack of comparative-anatomical preparations; the Bonn collection added pathological preparations, which had also been underrepresented in the eighteenth-century collections. The governors now needed to tell the rest of the world that their collections were up to scratch; the collections would not regain their fame if people continued to think they were inadequate. Leiden's governors used various channels to communicate their message.

First of all, the university's annual reports. These reports were sent to the Minister of Education, who used them to write the 'Report on the State of Education in the Netherlands'.[52] This report was sent to parliament and published in the *Staatscourant* (government gazette; the official publication containing laws and governmental announcements).[53] Usually, the universities also received a copy of the report. Hence, the contents of Leiden's annual reports mattered: their claims could potentially reach a much wider audience than just that of the Minister and his staff. Potential readers included politicians, governors at other universities, and, in the case of the *Staatscourant*, informed (and probably influential) members of the public – all of whom the Leiden governors were pleased to remind (or convince) of their university's top position. Indeed, the annual reports regularly stressed the high quality of their anatomical collections. For example, after the acquisition of the Brugmans collection, the governors wrote:

With regard to the acquisitions that this university made in the past
year, should in the first place be mentioned the very precious collection
of the late professor Brugmans, with which the university has acquired,
in particular in the field of comparative anatomy, a collection that is not
only able to compete with other collections of its kind in our fatherland,
but that may also exceed, in quality as well as in number, all other
collections of its kind, both within and beyond our fatherland; and
which honours the excellent talents of its previous owner (who, unfortunately for science, died before his time) just as much as it enlarges and
extends the fame and lustre of this university.[54]

The governors were thus not only claiming Leiden's comparative anatomy
collection was good; they were claiming that it was the best and, as such,
would boost the university's fame.

The annual reports might reach politicians and administrators, but they
would never be read outside the Netherlands. Yet, the governors wanted to
enjoy international fame as well. A collection catalogue would be an excellent means to this end, as curator Gerard Sandifort explained to the
governors:

It would be no less glorious for this university, if it were to become
widely known how the already renowned collection, consisting of the
individual cabinets of professors Rau, Albinus, van Doeveren and
others, has again been enlarged and become more suitable for teaching
all parts of anatomy with this [collection; the Brugmans collection].[55]

The governors, susceptible to Sandifort's arguments, decided to publish a
new catalogue.[56] Aimed at 'the learned world', it was written in Latin and
could therefore be read throughout Europe.[57] The catalogue described both
the Brugmans and the Bonn collections. In the preface, Sandifort wrote:
'The collection [of the Anatomical Cabinet] has been enriched and adapted
to the present-day state of science [*disciplinae*] ... Our museum has acquired
very important additions, for the collections of both Brugmans and Bonn
have been purchased.'[58] Brugmans's and Bonn's collections had 'adapted' the
university's anatomical collections 'to the present-day state of science'. Sandifort did not specify what this 'present-day state of science' entailed, but
this becomes clear from his descriptions of the new collections. Regarding
the Brugmans collection, he wrote: 'Brugmans ... left behind a collection of
preparations, by which comparative anatomy and pathology are elucidated
in many ways.'[59] And on the Bonn collection: 'Bonn's collection should be
praised no less, especially for its pathological section.'[60] The quotations
suggest that 'present-day state of science' meant having a sufficient number
of comparative and pathological anatomy preparations – precisely the
demand made by the RDHE.[61] The catalogue showed that Leiden's anatomical collections were up to date.

But other Dutch universities owned up-to-date collections as well, thanks to William I's donations. Utrecht had the Bleuland collection, rich in comparative anatomy; Groningen had the collections of Petrus Camper, Pieter de Riemer, and Gerbrand Bakker, all of high quality as well. The Leiden anatomical collections were no longer inadequate, but they were still not unique – much to the dismay of Leiden's governors, who did not wish to settle for anything less than excellence. It was not enough to comply with the decree's standards; the governors had to find a way to put Leiden *above* other universities, instead of next to them.

The second strategy: continue the past into the present

To distinguish themselves from the other universities, Leiden's governors used the university's glorious past as a claim to fame. The following quotation by the governors illustrates this strategy: 'It is known to Your Excellency [the Minister of Education] that Leiden University has been famous for over a century, mainly for the study of medicine, and that the fame Boerhaave acquired has continued to endure into our time.'[62] They suggested that the faculty and the collections were just as famous now as they had always been. The governors were trying to continue the past into the present.

In doing so, they used the past rhetorically – a common strategy in the Netherlands of the nineteenth century. As historian of science Nicholas Jardine explained in his analysis of the rhetoric of the laboratory revolution, if you want to use the past rhetorically, you can choose from two main strategies: normalization and dramatization.[63] Both serve to justify a practice or a state of affairs – laboratory-based medicine, for example, or Leiden's position as the first university of the Netherlands. Normalization justifies an aim or practice through presenting it as a natural development in a long tradition. Dramatization, on the other hand, does so through presenting it as a revolutionary break with the past. Leiden's governors used normalization, not dramatization: they justified Leiden's supposed status as the country's leading university by presenting it as the natural continuation of history.

But how does one do this – continue the past into the present? The first step is to adapt the past: you need to create an image of the past that resembles the image you want to create in the present. This may take some effort. For example, it took the Royal College of Surgeons in London years to position John Hunter as 'the first scientific surgeon' – a necessary step for the surgeons to be able to use Hunter's collections to position themselves as his heir, and hence, as scientific themselves, which in turn would make them more 'gentlemanly'.[64] In Leiden, however, creating the right image was easy. The governors needed an image that presented the university as a high-ranking institution with excellent anatomical collections. This was the standard image of the university in the eighteenth century, so the governors only had to remind their audience of that history; something that they did almost every time the governors mentioned the anatomical collections. The reminders were usually short and they often mentioned Albinus's name. In the quotation above, for example, the governors slipped in the name of Albinus when they explained to the Minister of Education that their collections did not comply with the standards of the RDHE. A similar strategy was used in the 1829–1830 report on the collections: 'The collection of anatomical preparations, with which the cabinets of Albinus, Brugmans and others have been placed, have been proved to meet with admiration from many local and foreign scholars constantly.'[65]

To continue the past into the present, however, it was not enough to recall past glory. Since Leiden's past glory was in the past, the governors needed to make a credible case that nothing had changed. They had to connect the past to the present – the second step in the rhetoric of normalization. The connection that the governors constructed started with a material link: the anatomical collections themselves. Obviously, the collections had a connection to the past, since the preparations *were from the past*. The argument ran as follows: the collections had been famous in the past, they continued to exist into the present, hence, their fame should continue to exist into the present as well. This connection was subsequently reinforced with other links: elements surrounding the collections – such as their curator or their catalogues – were tied to the past as well.

Some quotations from the annual reports demonstrate how the governors used the collections' curator to strengthen the connection to the past. As mentioned above, Gerard Sandifort was curator at the time the RDHE was issued. Sandifort had succeeded his father Eduard in 1799. The father-son relation was an excellent means to connect the nineteenth and eighteenth centuries. Consider the following phrase: '[the anatomical collections] being put under the special supervision of the virtuous son and worthy successor of the great Sandifort'.[66] The governors wrote this in 1819, when Gerard had been curator for 20 years. And yet, he was not called by his own name, but described as 'the virtuous son and worthy successor of the great Sandifort'. Eduard was a well-known curator and his collections were famous. By stressing that Gerard was his son, the governors were trying to associate that fame with their collections. This was strengthened by the addition 'worthy successor', which implied that Gerard had inherited his father's qualities. This suggestion can be found in other collection reports as well, for example: 'the praiseworthy professor Sandifort ..., who, on the heels of his worthy father, keeps the collection in the best condition'.[67]

Another means of linking past and present was the new collection catalogue, mentioned above. It was named *Museum Anatomicum Academiae Lugduno-Batavae, Volumen tertium*, to make clear that it was the sequel to *Museum Anatomicum Academiae Lugduno-Batavae, Volumen primum* and *Volumen secundum*, both published in 1793. The contents of the planned catalogue differed from the earlier catalogues, however. These had described all the preparations present in the collections, but the third volume would describe Brugmans's preparations. It would therefore have made sense to have presented it as a single collection catalogue, not as a sequel to the earlier museum catalogues. However, by doing this nevertheless, the governors were again linking the present to the past.

Eventually, the catalogue did cover both the Brugmans and the Bonn collections. This was contrary to the governors' original intentions, but the minister refused to pay for the catalogue if the Bonn collection was not included.[68] The governors may have intended to exclude the Bonn collection because it did not help establish a link with Leiden's past. Bonn was an anatomist in Amsterdam, and that was where he had built his collection. His collection was therefore associated with another city. Brugmans, on the other hand, was very much associated with eighteenth-century Leiden, where he had been a famous professor. This made his collection an excellent means to continue the past into the present.

Leiden thus distinguished itself from other universities by emphasizing its glorious past and reinforcing present links with the past through its collection, curator, and catalogues. This worked because, the rivalling collections in Utrecht and Groningen, unlike those in Leiden, did not embody a glorious past. The Camper collection in Groningen dated from the second half of the eighteenth century, which made it almost as old as the Albinus collection. However, although Camper was famous, Groningen University enjoyed little status nationally, let alone internationally, at the time. Whereas the Albinus collection permitted Leiden to associate itself with a period in which it had been 'the first of Europe', the Camper collection linked Groningen to a time when it had only been one of the four 'other' Dutch universities. The Bleuland collection in Utrecht was younger than both the Albinus and the Camper collections. It had been established during the French rule, one of the worst periods in the university's history. Utrecht University had almost ceased to exist – not exactly a period that the university wished to celebrate. Furthermore, neither Groningen nor Utrecht had owned significant anatomical collections before the RDHE had been issued; the collections acquired by Groningen and Utrecht were from the eighteenth century, but as *institutional* collections they were new. Leiden had owned anatomical collections before, making it easier to position the present-day anatomical collections as a continuation of the past.

The collections owned by Utrecht and Groningen did not offer these two universities a status-enhancing connection to the past, something of which they were well aware. Consider the following quotation from a letter from the Utrecht governors, in which they thanked the king for the Bleuland collection:

> We feel ourselves obliged to show Your Majesty our appreciation of and our great gratitude for this important and precious gift [the Bleuland collection], which, being a token of Your Royal generosity, will serve as a lasting ornament for this university and [which] will contribute, we believe, more than a little to its usefulness and flourishing. It

has even more value to this university, because it entirely lacked such a collection, and building such [a collection] would have taken a lot of time, effort, and money.[69]

The governors bluntly acknowledged that their university had lacked an anatomical collection. Instead of depicting the new collection as a continuation of the past, they portrayed it as a radical breach with that past. Leiden's governors, by contrast, presented the Brugmans collection as an addition to the existing collection. They saw their collection as a cumulative collection; and as its existence had been continuous over time, so should its status. Utrecht, on the other hand, described the Bleuland collection not as an addition or a continuation, but as a new beginning – Utrecht's governors were not normalizing, but dramatizing. They admitted that their anatomical collections had been useless in the past, but now, everything would change: the university would start to flourish.

The other Dutch universities did not use the history of their anatomical collections to enhance their present-day status. Beyond the Netherlands, however, several institutions used rhetorical strategies similar to those used by Leiden's governors. Cultural historian Rebecca Messbarger has written about anatomical collections in eighteenth-century Bologna.[70] The city's administrators, led by Archbishop (and future pope) Prospero Lambertini, aimed to restore the city's prestige by creating a new anatomy museum. The museum contained mainly wax models, newly made. Unlike the Albinus preparations, the collection itself was not historical, but it was explicitly intended to hark back to the public dissections for which Bologna had been famous in the seventeenth century. Although the collections themselves were not from the past, they did in a certain sense embody that past – and by presenting them as a continuation of the past, Bologna's administrators hoped to revive the city's former glory.

Another example can be found in nineteenth-century London, where the Royal College of Surgeons used the anatomical preparations of John Hunter to increase their status. When John Hunter died in 1793, he left behind one wife, two children, and over 13,000 preparations – which turned out to be an unfortunate combination. To provide money to take care of the collection, not to mention themselves, Hunter's family was forced to enter positions (his wife and son) and marriages (his daughter) that did not necessarily make them happy.[71] To their relief, in 1799 the British government paid them 15,000 pounds for Hunter's collection. Hunter's family was thus freed from the burden of his collection, but now the government had to find someone else willing to take care of it. The Company of Surgeons (the later Royal College of Surgeons) volunteered, in the hope that the collections would increase their prestige. As the College Council put it in the 1830s: 'The College derives an important accession to its scientific character from the possession of the Hunterian collection.'[72] Between 1799 and the 1830s, the surgeons had, as mentioned above, worked hard to turn Hunter into the father of scientific surgery, and they had subsequently used his collections to portray themselves as his sons. They suggested that they were simply continuing his work, for example by claiming that they were using Hunter's original arrangement.[73] In fact, they could not have been, because no-one really knew how Hunter thought his collection should be arranged.[74] But the College administrators and curators did not let this bother them: they focused on presenting the collection as a continuation of Hunter's work. That the College's museum is nowadays known as the Hunterian Museum, even though most of its preparations were not made by Hunter, shows how successful they were.

Preparations disconnected from their makers

Leiden University's governors thus combined two strategies to use their anatomical collections as a status symbol. First, they extended the collections to comply with the standards set in the RDHE and made sure that everybody knew about these extensions. Second, they suggested that nothing had changed since the eighteenth century. Although it seems contradictory to present the collections as both up to date and historical, in the first decades after the decree, the governors found it easy to do both. As the century progressed, however, the collections came to resist this dual meaning. Medical research and teaching kept changing, and the anatomical collections could remain up to date only if they changed as well, but in the process, they increasingly became separated from their past.

In Chapter 2, we saw how researchers reinterpreted individual preparations. Reinterpretation happened with all preparations, even those of the famous Albinus. An example can be found in the dissertation of the medical student Annee Leendert Erkelens, completed in 1902.[75] Erkelens investigated *retentio dentium*, the impaction of teeth. He used 18 preparations from the Leiden anatomical collections in his research, including a skull from the Albinus collection.[76] The skull showed a specific manifestation of the condition: teeth growing backwards. Two teeth in the upper jaw had grown upwards, with the roots below the teeth instead of above it. They can be seen in the stereographic photograph of the skull Erkelens included in his dissertation; see Figure 4.1.

The skull had been described and depicted before, by Albinus himself in his *Annotationes academicae* and by Eduard Sandifort in the first volume of the *Museum Anatomicum*.[77] Erkelens was aware of Albinus's description: he included it in his dissertation. (He probably knew about Sandifort's description, too, but he did not mention it.) However, for his research, Erkelens needed more information than that contained in the earlier descriptions of the preparation. In particular, he wanted exact measurements – such as the distance from the teeth to the body's median plane, which, Erkelens tells us, is 15 mm for the impacted tooth on the right and 10 mm for the tooth on the left.[78] Describing anatomies and pathologies with

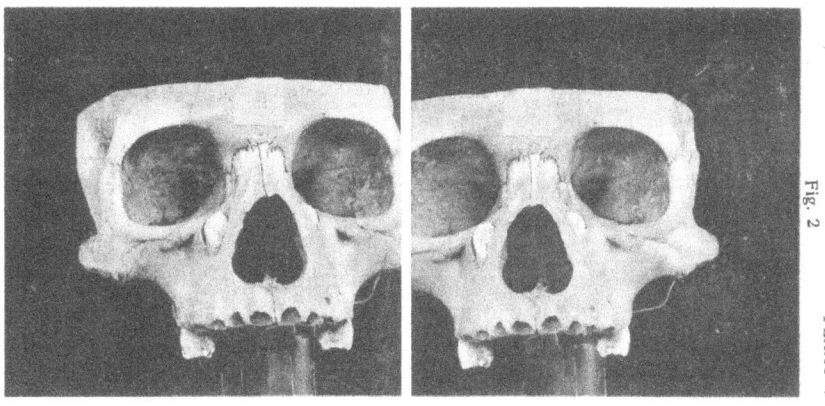

Figure 4.1 Skull collected by Bernhard Siegfried Albinus in the eighteenth century and reused by doctoral student Annee Leendert Erkelens around 1900. Stereographic photograph in Annee Leendert Erkelens, *Retentio dentium* (Leiden: IJdo, 1902). Courtesy of Leiden University Library, DISL 1902: 24.

such numerical precision had become common in the second half of the nineteenth century; it is therefore not surprising that Erkelens wanted exact figures, nor that Albinus and Sandifort did not provide them. Erkelens had to gather this information himself, and he encountered no problems in doing so: he simply reinvestigated the preparation. Not only did he take measurements, but he also tested whether the teeth could move (the one on the right did, the left one was almost immobile).

Students such as Erkelens, and researchers such as those discussed in Chapter 2, reinterpreted preparations to answer specific research questions. Collection curators also reinterpreted the collections, and they operated on a larger scale. The previous chapter described how, after the move in 1860, the collections as a whole were rearranged and reclassified. This changed all of the preparations. We saw how the preparations were detached from the (moral) stories that had made them interpretable to lay visitors; we will now learn how they were also disconnected from their makers, meaning that they could no longer be used by the governors to create a link with the past.

In the first half of the nineteenth century, users only had to read a preparation's label to find out who had made it. Curator Gerard Sandifort, who used the same labels as his father Eduard, included three pieces of information on the labels: the catalogue number, a description of the object, and the name of the maker (or, sometimes, the collector).[79] The catalogue number referred to the descriptions in the four volumes of the *Museum Anatomicum*. In these volumes, the Sandiforts described the collections from different makers (collectors) separately. A skull collected by Brugmans was described in the section on dry preparations in the Brugmans collection; a similar skull collected by Bonn was described in a different section, together with the other skulls from the Bonn collection. However, it is possible that both skulls were nonetheless placed next to each other on the Cabinet's shelves – we do not know the extent to which the classification system used in the catalogues was reflected in the actual arrangement of the preparations.

In his preface to the third volume of the *Museum Anatomicum* (1827), Gerard Sandifort seems to suggest that the collections were at least partly combined:

> When the Museum was enlarged so splendidly, it had to be rearranged and reordered; since it was made up of separate collections – those of Rau, Albinus, Van Doeveren, Ledeboer, Rocquette, Brugmans, and Bonn – it had to be given its own proper ordering and, as it were, face and character. And thus, I put together everything that had been separated until then and I made sure that, while everything ran according to an uninterrupted system, each preparation had a number and the name of the collection from which it was taken.[80]

The 'uninterrupted' system that Sandifort wrote about was not the classification system used in the catalogue – in that system, each collection had its own classification system, all employing different categories. But if the catalogue's system was not the 'uninterrupted' system, then this has to have been the system in which the preparations were arranged (*'disponendum'*). Yet, even in this 'uninterrupted' arrangement, Sandifort claimed, the individual collections remained recognizable.

Travel reports reveal that visitors did indeed distinguish preparations made by different anatomists. Take, for example, the account by the German doctor Wilhelm Horn, published in 1831. Horn offered a four-page list of objects visible in the Anatomical Cabinet, including

> Many vessel injections by Albinus – A single preparation by Ruysch, an injected child's head. Next, many preparations together, by Bonn, Brugmans, Sandifort, and Rau. – Injected organs of all kinds. – Stones, bladders, in particular by Van Doeveren: lymph-vessels, milts, livers; injected.[81]

Horn was able to identify the makers of the preparations. He suggested that he had seen a group of several injection preparations by Albinus; and that preparations made by Bonn, Brugmans, Sandifort, and Rau were also grouped together. Other visitor reports also regularly mentioned individual collections, showing that the visitors had at least learned that the Cabinet housed the collections of various anatomists.[82] We do not know whether these collections were kept strictly separate – probably not, considering Sandifort's remark. But even if they were mixed together to a certain extent, the connection between preparations and their makers was clear: in the catalogue, on the labels, and possibly (partly) in the actual arrangement.

After 1860, the clues that connected the preparations to their makers disappeared. As we have seen, curator Halbertsma took advantage of the move to the laboratory complex to rearrange the collections. The Albinus skull with the malformed teeth used by Erkelens ended up in room 11, cupboard KLM, in the section on 'human teratology', as one of '22 dry preparations (malformations of skull bones and teeth)'.[83] Albinus's name appeared neither on its label nor in its catalogue description. The individual collections were fully integrated, both in their actual arrangement and in the classification system.[84] Skulls were put with skulls; hearts with hearts; ears with ears – if they displayed the same body part, organ system or disease, preparations were put together, regardless of who had made them. The catalogues of Halbertsma (1860s) and Zaaijer (1892) did not even mention Albinus, Brugmans, Bonn, and the other Leiden anatomists.[85] Nor did the new labels on the individual preparations: they carried a description and a catalogue number, but no names of makers or collectors. Of course, some people still remembered that the collections contained preparations made by anatomists from previous centuries. From time to time, individual preparations were reconnected to their past. Zaaijer's inventory singles out a preparation made by seventeenth-century anatomist Frederik Ruysch, the only maker mentioned in the inventory.[86] And, as we have seen, Erkelens referred to Albinus when using his skull, but he did so in a purely medical fashion, not to stress the historical properties of the preparation. In general, it became difficult to discern the individuals behind the collections. In the early nineteenth century, all of the thousands of preparations on display had been directly connected to their makers; in the late nineteenth century, virtually all of these connections had disappeared.

The governors thus faced a problem: without a connection to the past, the collections could not function as a status symbol. Indeed, in the second half of the nineteenth century, the normalizing rhetoric disappeared from the governors' references to the anatomical collections. They still reported to the government on the collections (they were legally obliged to), but they never mentioned Albinus's name.[87] This does not mean, though, that the university ceased to draw on its past to enhance its present-day status. Consider, for example, what happened after the 1865 medical laws came into operation. Leiden felt somewhat threatened by these laws, which stressed the importance of practical teaching. In response, the university (more specifically, the medical professor Gerard Suringar) constructed an image of the famous Boerhaave as the first practitioner of bedside teaching, and then connected this image to the present. The suggestion was thereby made that theoretical medicine and practical teaching had long been, and would continue to be, combined at the Leiden medical faculty.[88]

The lost connection between preparations and their makers would later also pose a problem for medical historians. Owing to the rearrangement – and to the relabelling in particular – historians struggled to figure out who had made what. Understandably unhappy about all the tedious work they

had to undertake, they were keen to find someone to blame. Their eyes
landed on the collections' curators from the second half of the nineteenth
century: Hidde Halbertsma, Johannes Boogaard, and Teunis Zaaijer. In
1934, the authors of a historical article on the Leiden anatomical collections
stated the following:

> Lack of historical awareness, typical of the second half of the nineteenth
> century, and in addition lack of space in the institute in which anatomy was
> housed from 1859 until 1923 [the authors incorrectly date the 1860 move to
> 1859], resulted in the constant depletion of the contents of the old cabinets,
> which would not have happened had greater care been taken and had there
> been a greater inclination for these things. Part of the preparations were not
> only transferred to new jars or remounted, but, in these ahistorical times,
> old labels were also removed and all traces of the provenance of the pre-
> parations were destroyed. ... Because of these museological errors, the
> preparations lost their distinctive historical value.[89]

The authors, D. C. Geyskes and Cornelis J. van der Klaauw, accused the
three curators of 'museological errors' and claimed that these 'errors' stemmed from a lack of historical awareness, supposedly common in the second
half of the nineteenth century. But, as we saw above, the university had
continued to draw on its past, proving there was at least some sort of
awareness of history. Moreover, the three curators all valued the past in one
way or another. In his inaugural lecture, Zaaijer demonstrated that he was
well aware of the history of anatomy.[90] Boogaard, in turn, chaired the committee that erected a statue of Boerhaave. And Halbertsma treasured a
microscope made by Van Leeuwenhoek, on whose research he wrote his dissertation.[91] They were not ahistorical men. Yet, neither were they *primarily*
concerned with the historical value of the preparations. This is not, as
Geyskes and Van der Klaauw put it, a 'museological error'; on the contrary,
one could say. Halbertsma and his successors rearranged, reclassified, and
relabelled the preparations because they wanted to increase the preparations'
utility for what they considered the museum's primary purpose: teaching and
research.[92] They adapted the collections in line with changing medical practices and theories, something that was enabled by the preparations' potential
for reinterpretation.

The preparations were reused in research and teaching, and they were
arranged, classified, and labelled so as to best facilitate their new use.
Unfortunately for the governors (and for future medical historians), the
connection to the makers disappeared in the process. Since this connection
had been essential for Leiden's ability to distinguish itself from the other
Dutch universities, the governors stopped using the anatomical collections as
a status symbol. In the twentieth century, the connection was restored in
some cases, and part of the collections once again became a status symbol –
not for the university as a whole, but for the medical faculty specifically.

Leiden's anatomical collections in the twentieth century

In 1932, two men applied to the Leiden University Fund for money to clear out an old cabinet.[93] The men were J. A. J. Barge, anatomy professor, and C. A. Crommelin, director of the new Dutch historical science museum (Nederlandsch Historisch Natuurwetenschappelijk Museum, founded in 1931). The cabinet had formerly belonged to the Albinus brothers; it contained around 800 wet preparations from the 'old' Leiden anatomical collections. The preparations had been retrieved from the basement of the Anatomical Cabinet when the anatomy department moved to a new laboratory in 1923. It is unknown when, why, and by whom they were put in the basement, but it seems safe to assume that it was related to the lack of space in the museum rooms upstairs.[94] However, the fact that the preparations were moved to the basement does not mean they were considered useless – after all, they were *kept*, not thrown away, even though the glassware could probably have been put to good use elsewhere. Nonetheless, it seems likely that they were used less frequently than the preparations upstairs, especially considering the neglected condition in which they were found in 1923. They may have been stored for future use by researchers or students, much like the store preparations at the Royal College of Surgeons, which could remain in storage for decades until a new research question or technique made them relevant again.

Whatever the reason may have been that these preparations had ended up in the basement, once they were retrieved, they received quite some attention. In the course of two restoration projects, the majority of the wet preparations were reconnected with their makers. The first project took place in the 1930s, financed by the money Barge and Crommelin had requested from the Leiden University Fund. D. C. Geyskes, who worked as an assistant at the zoological laboratory, carried out most of the work. He was supervised by Cornelis van der Klaauw, deputy director of the Dutch historical science museum; together they wrote the historical article discussed above. The project aimed to catalogue the preparations and report on their condition. Geyskes and Van der Klaauw found 353 preparations carrying legible labels.[95] Most of the labels had been added by the Sandiforts, father and son, but they also discovered preparations with labels added by later nineteenth-century curators, suggesting that at least part of the preparations had spent some time in the Cabinet's museum upstairs before being moved to the basement.[96] In the end, they managed to match 271 preparations to a specific description in the *Museum Anatomicum*; and 17 to one of the collections described in the *Museum*, but not to a specific description. The preparations were returned to the Albinus cabinet and arranged according to collector. A conservation report was written, but no work was done on the preparations themselves (this had never been the intention of the project, probably because it would have cost too much time and money).

The preparations were in bad shape when they were taken out of the cabinet again, during the Second World War, to be moved – again – to the basement for safekeeping. Antonie Luyendijk-Elshout, later professor of medical history, described them as follows:

> Clearing out the mahogany cabinet resulted in a mournful spectacle. Eight hundred dirty jars, many of them with mouldy contents, had to be stored in the basement of the Anatomical Laboratory. Many preparations had gone dry; many old phials had cracked and were weather-stained. The corks had fallen into the jars; all that could [now] be seen of many beautiful intestine preparations was a turbid mass at the bottom of the cylindrical jars.[97]

After the War, Luyendijk-Elshout set to work: she restored preparations, topped them up, and relabelled them. She also created a new cataloguing system for the Anatomical Museum, which is still in use today. Furthermore, she painstakingly compared the preparations from the Albinus cabinet to the descriptions in the *Museum Anatomicum;* she matched 451 preparations, 180 more than Geyskes and Van der Klaauw. She also found 78 preparations described elsewhere (for example, in the Suringar catalogue). Nevertheless, 220 preparations remained disconnected from their makers – that is, 220 of the preparations in the Albinus cabinet, for many of the eighteenth-century preparations had never ended up in that cabinet. The *Museum Anatomicum* described almost 2,000 wet preparations, so some 1,200 must have ended up elsewhere. Some had no doubt been damaged or destroyed (for example, during the gunpowder disaster); some had been moved to the laboratories of physiology and pathological anatomy, and to the museum for natural history; and some remained tucked away in the other collections in the anatomical laboratory. Geyskes and Van der Klaauw wrote: 'Without a doubt, many preparations in the new section of the collection of the new Anatomical Institute stem from the old cabinets. It is virtually impossible to find out for sure.'[98]

A similar picture holds true for the dry preparations: they were completely absent from the Albinus cabinet, yet abundant in the *Museum Anatomicum.* In the second half of the twentieth century, when the entire Anatomical Museum was catalogued (mostly by Elshout), some dry preparations were also reconnected to their makers. Some of the dry preparations were entered in the catalogue as 'from unknown origin', but others could be linked to their makers. Often, the makers' names had been written *on* the preparations, solving the problem of labels becoming illegible or getting lost. Yet even writing on preparations offers no guarantees, as historian Flavio Häner's microhistory of an early modern skull in the Basel Anatomy Museum shows. In 1971, all historical inscriptions were removed from the skull, to make it fit a new arrangement based on contemporary anatomical science. In this way, the skull was detached from its history.[99]

In the second half of the twentieth century, some of the eighteenth-century preparations were put on display in Museum Boerhaave, the successor to the Dutch Historical Science Museum. But most of them remained in the medical faculty's Anatomical Museum, where they can still be found. And, as happened 200 years ago, the preparations are used to create a status-enhancing link to Leiden's glorious past – with Albinus once more taking centre stage. He greets us outside the building: next to the back entrance, above the bicycle racks, we see a gigantic poster of an engraving from Albinus's famous anatomical atlas *Tabulae sceleti et musculorum corporis humani*. It has a Seneca quotation as its caption: *Non scholae sed vitae discimus* (We do not learn just for school, but for life) (Figure 4.2).

Inside, we find Albinus's old cabinet – like the nineteenth-century university governors, the twenty-first-century medical administrators use not only the preparations themselves, but also elements surrounding the collections. On the wall adjacent to the cabinet, we find portraits of famous Leiden anatomists. And then, of course, there are the preparations themselves: Albinus's, Bonn's, Brugmans's – all reminding us of Leiden's glorious past. It is almost as though history were repeating itself, but with two major differences, both consequences of the prolonged use of the anatomical collections. First, these days, the old collections are a status symbol for the medical centre, not the

Figure 4.2 Back entrance to Leiden University Medical Center's teaching building. An illustration taken from Bernhard Siegfried Albinus's anatomical atlas reminds visitors of the medical faculty's glorious past. The illustrations in Albinus's atlas were made by Jan Wandelaar, who is not credited on this poster. Photograph by the author, 2012.

university as a whole. This is because the collections retreated into the medical faculty in the second half of the nineteenth century; they lie out of the university administrators' reach, accessible to the administrators in the medical centre only. And second, in the nineteenth century, all of the thousands of preparations on display connected the present to the past. In the twenty-first century, this number has dwindled to a few hundred; the other historical preparations have had to bid a final adieu to their makers.

Acknowledgements

Parts of this chapter have been published in an earlier version in my book chapter 'Adieu Albinus: How the preparations in the nineteenth-century Leiden anatomical collections lost their past', in *The Fate of Anatomical Collections*, edited by Rina Knoeff and Robert Zwijnenberg, 113–28. The History of Medicine in Context (Farnham: Ashgate, 2015).

Notes

1 On the history of the College and its collections, see Alberti, 'Organic Museum'; Blandy and Lumley, *Royal College*; Chaplin, 'John Hunter'.
2 Aerts, 'Staat in verbouwing', 47–48.
3 De Rooy, *Republiek van rivaliteiten*, 15–36.
4 On William I's policy of unification, see Vosters and Weijermars, *Taal, cultuurbeleid en natievorming*.
5 Luiten van Zanden and Riel, *Nederland 1780–1914*, 206.
6 On William I's economic policy, see Luiten van Zanden and Riel, 109–203. On his educational policy, see De Wolf, 'Bibliografie' (on education in general); Groen, *Onderwijs*; Roelevink, 'Eenen eik' (on the universities in particular). On his cultural policy, including his use of museums as unifying instruments, see Pots, *Cultuur, koningen en democraten*, 59–84.
7 De Jaucourt, 'Leyde, Lugdunum Batavorum', 451.
8 F. B. Albinus and Eduard Sandifort, 'Rapport over het kabinet van Albinus', 7 November 1771, cited in Molhuysen, *Bronnen*, 6: 18*.
9 Zoeteman, *Studentenpopulatie*, 186–88, 252–53.
10 Otterspeer, *Werken van de wetenschap*, 209.
11 Knappert, *Ramp van Leiden*, 23.
12 Van der Burg, 'Nederland onder Franse invloed', 107.
13 Otterspeer, *Werken van de wetenschap*, 210.
14 Minutes of the governors, 1 February 1800, cited in Molhuysen, *Bronnen*, 7: 131.
15 Otterspeer, *Werken van de wetenschap*, 210.
16 Minutes of the governors, 4 February 1807, cited in Molhuysen, *Bronnen*, 7: 312–13.
17 Cited without a source in Kernkamp, *Utrechtsche academie*, 379.
18 Jamin, *Kennis als opdracht*, 104.
19 Visser, 'Jan Bleuland', 51; Blanken, *Aantal studenten*, 79.
20 Jensma and De Vries, *Veranderingen*, 79–80; Otterspeer, *Werken van de wetenschap*, 75–77; Roelevink, 'Rapport', 13.
21 Sluijter, '*Tot ciraet*', 59–61; Aerts, 'Staat in verbouwing', 52–53.
22 Cuvier and Noël, *Rapport*.
23 Meerman, *Commissie*; Van Swinden et al., *Vertoog over de universiteiten*.
24 Roelevink, 'Rapport'.

25 In 1816, a similar Decree on Higher Education was issued for the Southern Netherlands. Again, three universities were established: Ghent, Leuven, and Liège. These universities formed part of the Netherlands until 1830, when Belgium seceded – William I's attempt to unite the North and the South had not been very successful. During the 14-year period that the southern universities lay in the Netherlands, the northern universities, including Leiden, paid little attention to them. That is why I have left them out of the discussion here; Leiden had no concerns about losing its status to these universities.

26 See for example Otterspeer, *Wiekslag*, 5, and Calkoen, 'Onder studenten', 190. I am not the first to point out that the RDHE ranked all three universities equally; see Van Berkel, *Voetspoor*, 103–4.

27 Roelevink, 'Rapport', 26–27; De Geer van Jutphaas, 'Regeling van het hooger onderwijs', 232–33.

28 De Geer van Jutphaas, 'Regeling van het hooger onderwijs', 216; Otterspeer, 'Vereenvoudiging en bezuiniging', 244.

29 De Geer van Jutphaas, 'Regeling van het hooger onderwijs', 220.

30 'Organiek Besluit Hooger Onderwijs', 2 August 1815 (hereafter cited as RDHE 1815), section 'vijfde titel'.

31 RDHE 1815, articles 169, 177, 180.

32 RDHE 1815, article 177.

33 Annual report of the university 1815–1816, file 270, Archief van Curatoren 1815–1877 (hereafter cited as AC2), Leiden University Library.

34 Sandifort, *Museum Anatomicum 1*; Sandifort, *Museum Anatomicum 2*.

35 It is hard to say exactly how many preparations there were: some descriptions cover more than one preparation, and some preparations are listed twice. The numbers should thus be taken with some caution, but at least they give a rough indication.

36 Sandifort, *Museum Anatomicum 1*, 66–69, 88–90.

37 Sandifort, 111. On Van Doeveren's animal preparations, see also Hendriksen, *Elegant Anatomy*, 110–14. Note that it is unclear how many of Van Doeveren's preparations were still present in 1815, as the collection had been damaged in the gunpowder disaster.

38 This does not mean that private collections disappeared, or that 'private' and 'institutional' were the only two useful categories; a wide range of ownership constructions existed, all of them in use throughout the nineteenth century. See Alberti, 'Owning and Collecting'.

39 Van Swinden et al., *Vertoog over de universiteiten*, 118–19.

40 Van der Hoeven, *Berigt*, 16.

41 Van der Knaap, 'Jan Bleuland', 13–17.

42 Haneveld, 'Koning Willem'.

43 The university administration was regulated in the RDHE (RDHE 1815, section 'zevende titel'). On the actual administrative practices resulting from the decree, see Dorsman, 'Slaapmutsen en ornamenten?'

44 Otterspeer, *Werken van de wetenschap*, 77. For a full list of nineteenth-century governors in Leiden and at the other universities, see Jensma and De Vries, *Veranderingen*, 75–126.

45 Governors to Minister of Education, 17 January 1822, file 228, document 5, AC2.

46 RDHE 1815, article 234.

47 See for example the Annual report on the university collections 1821–1822, file 228, document 90, AC2; governors to Minister of Education, 28 January 1823, file 229, document 9, AC2; Annual report of the university 1822–1823, file 229, document 57, AC2.

48 It is often stated that the Bonn collection was acquired in 1819, just like the Brugmans collection. (See for example Elshout, *Leidse kabinet*, 88; Museum Boerhaave, *Leidse anatomie*, 6–7.) However, the university archives clearly show

that this did not happen until 1822; see for example the letter of the Minister of Education to the governors, 22 October 1822, file 76, document 162, AC2.

49 Minister of Education to governors, 22 October 1822, file 76, document 162, AC2.

50 Sandifort, 'Rapport aan curatoren over het Museum Anatomicum Andreae Bonn, voor het Theatrum Anatomicum der Leidsche Hoogeschool aangekocht', 1823, file 1807, Bibliotheca Publica Latina Collection, Leiden University Library; Sandifort to governors, 21 March 1823, file 77, document 40, AC2.

51 The Leiden part of the collection is catalogued in Sandifort, *Museum Anatomicum 3*.

52 The full titles of these reports vary; between 1816 and 1857–58, they were called *Verslag nopens den staat der hooge, middelbare en lagere scholen in het Koningrijk der Nederlanden*.

53 Jensma and Vries, *Veranderingen*, 54.

54 Annual report of the university 1819, 9 January 1820, file 226, document 3, AC2.

55 Sandifort to governors, 21 January 1823, file 77, document 10, AC2.

56 Governors to the Minister of Education, 28 January 1823, file 229, document 9, AC2.

57 Sandifort, *Museum Anatomicum 3*, Praefatio, 4.

58 Sandifort, Praefatio, 3.

59 Sandifort, Praefatio, 3–4.

60 Sandifort, Praefatio, 5.

61 Of course, an international audience would not have known about the RDHE and its demands, but they would have known that comparative and pathological anatomy had become important disciplines in medicine, and that a proper anatomical collection contained preparations from both fields. Furthermore, the catalogues were read inside the Netherlands as well; parts of the preface may have been intended mainly for a national audience.

62 Governors to Minister of Education, 17 January 1822, file 228, document 5, AC2.

63 Jardine, 'Laboratory Revolution in Medicine', 314.

64 Jacyna, 'Images of John Hunter' shows how the College used the annual Hunterian Orations to turn Hunter into the first scientific surgeon.

65 Annual report of the university 1829–1830, file 270, AC2.

66 Annual report of the university 1817–1818, 8 January 1819, file 226, document 4, AC2.

67 Annual report of the university 1819, 9 January 1820, file 226, document 3, AC2.

68 Sandifort to governors, 11 May 1823, file 77, document 63, AC2.

69 Governors to King, 4 November 1816, file 37, document 320, Archief van het College van Curatoren van de Rijksuniversiteit te Utrecht, Utrechts Archief, 59, Utrecht.

70 Messbarger, *The Lady Anatomist*, 1–51.

71 Chaplin, 'John Hunter', 259.

72 Council Minutes, 22 August 1833, Minute Book Vol 6, 1833–1838, file 2/1/2, College Governance Fonds, Royal College of Surgeons Archives, RCS-GOV, London.

73 See for example Royal College of Surgeons, *Synopsis*, 3.

74 Cunningham, 'Quis Custodiet Ipsos Custodes?'

75 Erkelens, *Retentio dentium*.

76 Preparation Ab0189 in the present-day museum database.

77 Albinus, *Academicarum annotationum*, 1: 54–55, 90–91; Sandifort, *Museum Anatomicum 1*, 86, entry CCCXLVII.

78 Erkelens, *Retentio dentium*, 10.

79 Elshout, *Leidse kabinet*, 11. The Sandiforts used the name of the anatomist who had built the collection. Usually, this anatomist was both the maker and the collector of the individual preparations; in the eighteenth century, anatomists tended to make their own preparations. This certainly applies to the Albinus collection. In some of the other collections, such as the Brugmans collection, not all preparations were made by the collection builder, meaning that strictly

speaking, some of the preparations were connected not to their maker, but to their collector – but either way, they were connected to their past. Because most preparations were connected to their makers, I use the term 'maker' instead of 'maker or collector'.

80 Sandifort, *Museum Anatomicum 3*, Praefatio, 4.

81 Horn, *Reise*, 1: 360.

82 See for example Guislain, *Lettre médicale sur la Hollande*, 91; MacGregor, *My Note Book*, 168; Van Meerten, *Reis*, 304.

83 Teunis Zaaijer, 'Inventaris der verzameling in het Anatomisch Kabinet van de Rijks Universiteit te Leiden', 1892, p. 31, archives Anatomisch Museum (no inventory number), Leiden University Medical Center.

84 The 1892 inventory by Zaaijer, discussed in the previous chapter, listed the preparations by cupboard, and shows that the classification system and the arrangement coincided.

85 Teunis Zaaijer, 'Inventaris der verzameling in het Anatomisch Kabinet van de Rijks Universiteit te Leiden', 1892, archives Anatomisch Museum (no inventory number), Leiden University Medical Center. The classification system used by Halbertsma is described in Elshout, *Leidse kabinet*, 11. Until recently, Halbertsma's catalogue was in the archives of the Leiden Anatomical Museum, but its present location is unfortunately unknown.

86 Teunis Zaaijer, 'Inventaris der verzameling in het Anatomisch Kabinet van de Rijks Universiteit te Leiden', 1892, p. 6, archives Anatomisch Museum (no inventory number), Leiden University Medical Center.

87 The annual reports from the second half of the nineteenth century can be found in files 271–73, AC2; and in files 1552–59, Archief van Curatoren 1878–1953, Leiden University Library.

88 Knoeff, 'Boerhaave at Leiden', 269–79; see also Suringar, 'Leidsche geneeskundige faculteit'; Suringar, 'Theoretisch-geneeskundig onderwijs'.

89 Geyskes and Van der Klaauw, 'Der heutige Zustand', 181–82.

90 Zaaijer, *Ontleedkundige techniek*.

91 Johann Czermák, who visited the Leiden collections in 1850, described how Halbertsma showed him the Leeuwenhoek microscope (Czermák, *Gesammelte Schriften*, 1: 174). For Halbertsma's dissertation: Halbertsma, *Dissertatio*.

92 I write 'museum' because Geyskes and Van der Klaauw used that word, but 'collections' would be more suitable here: the Cabinet's preparations, of course, were not just for display in the museum, but also for handling in other research and teaching spaces.

93 Elshout, *Leidse kabinet*, 2; Geyskes and Van der Klaauw, 'Der heutige Zustand', 182–83. Note that in this section, I use the ambiguous word 'cabinet' instead of the clearer 'cupboard'; I do this because of the historical connotation of the word cabinet. As this section shows, the historical character of the cupboard intended here is pivotal.

94 This is also suggested by Geyskes and Van der Klaauw, 'Der heutige Zustand', 182.

95 For a detailed description of the results, see Geyskes and Van der Klaauw, 'Der heutige Zustand'.

96 Even the preparations with the Sandifort labels were not necessarily put in the basement immediately after the move; the reclassification of the collection was only completed at the end of the nineteenth century. Until then, the museum probably still contained preparations with the old labels. Elshout wrote she had found at least eight different types of labels, several of them from the second half of the nineteenth century, and some from an exhibition held in 1915 (Elshout, *Leidse kabinet*, 11).

97 Elshout, 3.

98 Geyskes and Van der Klaauw, 'Der heutige Zustand', 182.

99 Häner, 'Restoration Reconsidered'.

Bibliography

Manuscript sources

Leiden University Library, Special Collections:
Archief van Curatoren 1815–1877;
Archief van Curatoren 1878–1953;
Bibliotheca Publica Latina Collection.
Leiden University Medical Center:
Archives Anatomisch Museum.
Utrecht, Utrechts Archief:
59, Archief van het College van Curatoren van de Rijksuniversiteit te Utrecht.
London, Royal College of Surgeons Archives:
RCS-GOV, College Governance Fonds.

Printed sources

Aerts, Remieg. 'Een staat in verbouwing: Van republiek tot constitutioneel konink-rijk, 1780–1848'. In *Land van kleine gebaren: Een politieke geschiedenis van Nederland 1780–1990*, by Remieg Aerts, Herman de Liagre Böhl, Piet de Rooy, and Henk te Velde, 11–95, 4th ed. Nymegen: SUN, 2004.

Alberti, Samuel J. M. M. 'Owning and Collecting Natural Objects in Nineteenth-Century Britain'. In *From Private to Public: Natural Collections and Museums*, edited by Marco Beretta, 141–54. Sagamore Beach: Science History Publications, 2005.

Alberti, Samuel J. M. M. 'The Organic Museum: The Hunterian and Other Collections at the Royal College of Surgeons of England'. In *Medical Museums: Past, Present, Future*, edited by Samuel J. M. M. Alberti and Elizabeth Hallam, 17–29. London: Royal College of Surgeons of England, 2013.

Albinus, Bernhard Siegfried. *Academicarum annotationum*. Vol. 1. Leiden: Verbeek, 1754.

Berkel, Klaas van. *In het voetspoor van Stevin: Geschiedenis van de nat-uurwetenschap in Nederland, 1580–1940*. Meppel: Boom, 1985.

Blandy, John P., and John S. P. Lumley, eds. *The Royal College of Surgeons of England: 200 Years of History at the Millennium*. London: Royal College of Surgeons of England, 2000.

Blanken, G. H. *Het aantal studenten aan de Hoogeschool te Leiden van 1775 tot en met 1868*. Leiden: Steenhoff, 1869.

Burg, Martijn Jacob van der. 'Nederland onder Franse invloed: Cultuurtransfer en staatsvorming in de Napoleontische tijd, 1799–1813'. PhD diss., University of Amsterdam, 2007. http://hdl.handle.net/11245/1.272472.

Calkoen, Godert Theodoor Allard. 'Onder studenten: Leidse aanstaande medici en de metamorfose van de geneeskunde in de negentiende eeuw (1838–1888)'. PhD diss., Leiden University, 2012. https://openaccess.leidenuniv.nl/handle/1887/20129.

Chaplin, Simon. 'John Hunter and the "Museum Oeconomy", 1750–1800'. PhD diss., University of London, 2009.

Cunningham, Andrew. 'Quis Custodiet Ipsos Custodes? Or, What Richard Owen Did to John Hunter's Collection'. In *The Fate of Anatomical Collections*, edited by Rina Knoeff and Robert Zwijnenberg, 23–52. The History of Medicine in Context. Farnham: Ashgate, 2015.

Cuvier, Georges, and J. F. M. Noël. *Rapport sur les établissemens d'instruction publique en Hollande et sur les moyens de les réunir à l'Université Impériale.* Paris: Fain, [1811].

Czermák, Johann. *Gesammelte Schriften.* Vol. 1. Leipzig: Engelmann, 1879.

Dorsman, Leen. 'Slaapmutsen en ornamenten? Over het bestuur van een universiteit in de negentiende eeuw'. *Studium* 1(2008): 32–46.

Elshout, Antonie M. *Het Leidse kabinet der anatomie uit de achttiende eeuw: De betekenis van een wetenschappelijke collectie als cultuurhistorisch monument.* Leiden: Universitaire Pers Leiden, 1952.

Erkelens, Annee Leendert. *Retentio dentium.* Leiden: IJdo, 1902.

Geer van Jutphaas, B. J. L. de. 'De regeling van het hooger onderwijs in Nederland in 1814'. *Nieuwe bijdragen voor regtsgeleerdheid en wetgeving* 19(1869): 212–73.

Geyskes, D. C., and Cornelis J. van der Klaauw. 'Der heutige Zustand der anatomischen Kabinette früherer Jahrhunderte in Leiden'. *Janus: Archives internationales pour l'histoire de la médecine et pour la géographie médicale* 38(1934): 179–92.

Groen, Marten. *Het wetenschappelijk onderwijs in Nederland van 1815 tot 1980: Een onderwijskundig overzicht.* 3 vols. Eindhoven: Technische Hogeschool Eindhoven, 1987–1989.

Guislain, Joseph. *Lettre médicale sur la Hollande, adressé à MM les membres de la Société de Médecine de Gand.* Ghent: Gyselynck, 1842.

Halbertsma, Hidde Justusz. *Dissertatio historico-medica inauguralis: De Antonii Leeuwenhoeckii meritis in quasdam partes anatomiae microscopicae.* Deventer: De Lange, 1843.

Häner, Flavio. 'Restoration Reconsidered: The Case of Skull Number 1-1-2/27 at the Anatomy Museum of the University of Basel'. In *The Fate of Anatomical Collections*, edited by Rina Knoeff and Robert Zwijnenberg, 247–62. The History of Medicine in Context. Farnham: Ashgate, 2015.

Haneveld, G. T. 'Koning Willem de Eerste en de anatomische collecties'. In *Acta octavi conventus historiae scientiae medicinae matheseos naturalumque excolendae Bergae ad Zomam* 2: 83–90. Amsterdam: Meesters, 1978.

Hendriksen, Marieke M. A. *Elegant Anatomy: The Eighteenth-Century Leiden Anatomical Collections.* History of Science and Medicine Library 47. Leiden: Brill, 2015.

Hoeven, Jan van der. *Berigt omtrent het mij verleende ontslag als opperdirecteur van 's Rijks Museum van Natuurlijke Historie te Leiden.* Amsterdam: J. H. Gebhard, 1860.

Horn, Wilhelm. *Reise durch Deutschland, Ungarn, Holland, Italien, Frankreich, Großbritannien und Irland; in Rücksicht auf medicinische und naturwissenschaftliche Institute, Armenpflege u. s. w.* Vol. 1. Berlin: Enslin, 1831.

Jacyna, L. Stephen. 'Images of John Hunter in the Nineteenth Century'. *History of Science* 21(1983): 85–108.

Jamin, Hervé. *Kennis als opdracht: De Universiteit Utrecht, 1636–2001.* Utrecht: Matrijs, 2001.

Jardine, Nicholas. 'The Laboratory Revolution in Medicine as Rhetorical and Aesthetic Accomplishment'. In *The Laboratory Revolution in Medicine*, edited by Andrew Cunningham and Perry Williams, 304–23. Cambridge: Cambridge University Press, 1992.

Jaucourt, Louis de. 'Leyde, Lugdunum Batavorum'. In *Encyclopédie, ou dictionnaire raisonné des sciences, des arts et des métiers, par une société de gens de lettres*, edited by Denis Diderot and Jean le Rond d'Alembert, 9: 451–52. Paris, 1765.

University of Chicago: *ARTFL Encyclopédie Project* (Autumn 2017 Edition), edited by Robert Morrissey and Glenn Roe. http://encyclopedie.uchicago.edu/.

Jensma, Goffe, and H. de Vries. *Veranderingen in het hoger onderwijs in Nederland tussen 1815 en 1940*. Hilversum: Verloren, 1997.

Kernkamp, Gerhard Willem. *De Utrechtsche academie, 1636–1815*. Vol. 1 of *De Utrechtsche Universiteit, 1636–1936*. Utrecht: Oosthoek, 1936.

Knaap, Emilie C. van der. 'Prof. dr. Jan Bleuland (1756–1838) en het Museum Bleulandium'. Master's thesis, Utrecht University, [2001].

Knappert, Laurentius. *De ramp van Leiden, 12 januari 1807, na honderd jaar herdacht*. Schoonhoven: Van Nooten, 1906.

Knoeff, Rina. 'Boerhaave at Leiden: Communis Europae Praeceptor'. In *Centres of Medical Excellence? Medical Travel and Education in Europe, 1500–1789*, edited by Ole Peter Grell, Andrew Cunningham, and Jon Arrizabalaga, 269–86. The History of Medicine in Context. Farnham: Ashgate, 2010.

Luiten van Zanden, Jan, and Arthur van Riel. *Nederland 1780–1914: Staat, instituties en economische ontwikkeling*. Amsterdam: Balans, 2000.

MacGregor, John. *My Note Book*. Vol. 1. London: John Macrone, 1835.

Meerman, Johan. *De commissie tot de formatie der Openbare en Koninklijke Hooge schoolen, en de aanmoediging van de wetenschappen en der geleerden aan Zijne Majesteit den Koning*. The Hague, 1807.

Meerten, A. B. van. *Reis door het Koningrijk der Nederlanden en het Groothertogdom Luxemburg, voor jonge lieden*. Vol. 5. Amsterdam: Schalekamp en Van de Grampel, 1829.

Messbarger, Rebecca. *The Lady Anatomist: The Life and Work of Anna Morandi Manzolini*. Chicago: University of Chicago Press, 2010.

Molhuysen, P. C. *Bronnen tot de geschiedenis der Leidsche Universiteit*. 7 vols. The Hague: Nijhoff, 1913–1924.

Museum Boerhaave. *Leidse anatomie in Museum Boerhaave*. Leiden: Museum Boerhaave, 2000.

Otterspeer, Willem. 'Vereenvoudiging en bezuiniging: Een negentiende-eeuwse discussie over taakverdeling en concentratie'. In *Universiteit in beweging: Een aantal beschouwingen bij gelegenheid van het 410-jarig bestaan van de Rijksuniversiteit te Leiden*, edited by Fr. van der Meer, 239–61. Leiden: Rijksuniversiteit Leiden, 1985.

Otterspeer, Willem. *De wiekslag van hun geest: De Leidse universiteit in de negentiende eeuw*. Hollandse historische reeks 18. The Hague: Stichting Hollandse Historische Reeks, 1992.

Otterspeer, Willem. *De werken van de wetenschap: De Leidse universiteit, 1776–1876*. Amsterdam: Bert Bakker, 2005.

Pots, Roel. *Cultuur, koningen en democraten: Overheid & cultuur in Nederland*. Nijmegen: SUN, 2000.

Roelevink, J. '"Eenen eik, die hondert jaren behoefde, om groot te worden": Koning Willem I en de universiteiten van het Verenigd Koninkrijk'. In *Staats- en natievorming in Willem I's koninkrijk, 1815–1830*, edited by C. A. Tamse and Els Witte, 286–309. Brussel: VUB-Press, 1992.

Roelevink, J. 'Het rapport van de commissie Van der Duyn van Maasdam over het hoger onderwijs uit 1814'. *Batavia academica: Bulletin van de Nederlandse Werkgroep Universiteitsgeschiedenis* 10(1992–1993): 1–61.

Rooy, Piet de. *Republiek van rivaliteiten: Nederland sinds 1813*. Amsterdam: Mets & Schilt, 2002.

Royal College of Surgeons. *Synopsis of the Arrangement of the Preparations in the Gallery of the Museum of the Royal College of Surgeons, for the Use of Visitors.* London, 1818.

Sandifort, Eduard. *Museum Anatomicum Academiae Lugduno-Batavae, Volumen primum.* Leiden: Luchtmans, 1793.

Sandifort, Eduard *Museum Anatomicum Academiae Lugduno-Batavae, Volumen secundum.* Leiden: Luchtmans, 1793.

Sandifort, Gerard. *Museum Anatomicum Academiae Lugduno-Batavae, Volumen tertium.* Leiden: Luchtmans, 1827.

Sluijter, Ronald. *'Tot ciraet, vermeerderinge ende heerlyckmaeckinge der universiteyt': Bestuur, instellingen, personeel en financiën van de Leidse universiteit, 1575–1815.* Hilversum: Verloren, 2004.

Suringar, Gerard Conrad Bernard. 'De Leidsche geneeskundige faculteit in het begin der achttiende eeuw: Boerhaave en zijne ambtgenooten'. *Nederlands tijdschrift voor geneeskunde* 10(1866): 1–39.

Suringar, Gerard Conrad Bernard. 'Het theoretisch-geneeskundig onderwijs van Boerhaave: De klinische lessen door hem en zijn ambtgenoot Herman Oosterdijk Schacht gegeven'. *Nederlands tijdschrift voor geneeskunde* 10(1866): 199–225.

Swinden, J. H. van, J. A. Bennet, Joh. Valckenaer, and J. F. van Beeck Calkoen. *Vertoog over de universiteiten, met betrekking tot het stelsel van openbaar onderwijs, en tot alle de inrigtingen die tot hetzelve, middellijk of onmiddellijk behooren.* Amsterdam, 1809.

Visser, Robert Paul Willem. 'Jan Bleuland (1756–1838): Hartstochtelijk verzamelaar'. In *Zes keer zestig: 360 jaar universitaire geschiedenis in zes biografieën*, edited by Hervé Jamin, 47–55. Utrecht: Universiteit Utrecht, 1996.

Vosters, Rik, and Janneke Weijermars, eds. *Taal, cultuurbeleid en natievorming onder Willem I.* Brussel: Paleis der Academiën, 2011.

Wolf, H. C. de. 'Bibliografie: Onderwijs en opvoeding in de Noordelijke Nederlanden'. In *Algemene geschiedenis der Nederlanden*, edited by D. P. Blok, 11: 392–93. Weesp: Fibula-Van Dishoeck, 1983.

Zaaijer, Teunis. *Het gewigt eener doelmatige ontleedkundige techniek.* Leiden: Hazenberg, 1866.

Zoeteman, Martine. *De studentenpopulatie van de Leidse universiteit, 1575–1812: 'Een volk op zyn Siams gekleet eenige mylen van Den Haag woonende'.* Leiden: Leiden University Press, 2011. https://openaccess.leidenuniv.nl/handle/1887/16453.

Conclusion
In perpetual motion, anatomical collections then and now

We have seen how nineteenth-century students, researchers, lay visitors, and administrators handled anatomical collections, in particular at Leiden University. Let us now turn to the final audience: twenty-first-century historians. How have I used nineteenth-century collections, and what might other historians gain from this use? And, apart from morbid anecdotes, does my story offer anything to non-historians? What does this book contribute to our understanding of anatomical collections, both historical and contemporary?

The four chapters above explained how different audiences used anatomical collections in the nineteenth century. The chapter on students demonstrated how anatomical preparations were not just looked at in the museum but actively handled in all medical teaching spaces. In the second chapter, we discovered that researchers also used preparations lids-off and hands-on. Moreover, as the nineteenth-century afterlife of the Brugmans collection showed, researchers continued to use old collections in new medical practices. We saw how this continuous reinterpretation followed on from a special feature of preparations: they are made of what they represent. The chapter on lay visitors explained how and why anatomical collections, unlike many other types of collections in the nineteenth century, were not opened up to the lay public, but instead ended up in medical laboratories, where they were hard to approach and interpret unless you had a medical background. And finally, the chapter on institutional administrators discussed how reinterpretation disconnected the anatomical preparations from their past, much to the regret of the Leiden university governors, who were no longer able to use the anatomical collections as a status symbol.

Collected together, the four chapters build a book: this book. And I would like to think that this book, like a proper collection, is more than the sum of its parts; that, if read from beginning to end, it allows its audience – that would be you – to acquire a knowledge of nineteenth-century anatomical collections that transcends the insights offered in the individual chapters. I would summarize this knowledge in two sentences. First, in the nineteenth century, medical audiences continued to use anatomical collections and non-medical audiences stopped using them. And second, these developments are causally related to each other and to the specific properties of anatomical preparations. To highlight this causality, let me now weave the individual chapters into one story.

If we take the Leiden anatomical collections as the story's protagonist, then their moving to the laboratory complex becomes its turning point. After the move, which took place in 1860, non-medical audiences could no longer use the collections as freely as they had done in the past. Lay visitors struggled to even enter the new Cabinet, with its distant location, unwelcoming building, and closed atmosphere. And inside the Cabinet, the new arrangement stripped away from the preparations the stories of unhappy marriages, crimes committed, and famous giants, and replaced them with 'scientific' anatomical and pathological descriptions. In addition to losing their stories, the preparations were detached from their makers. The university governors could no longer use them to connect the university's present to its glorious past. Anatomical preparations resist historization, a disadvantage for lay visitors as well, because historization helps scientific objects to move into public museums.

What caused this move and rearrangement, which closed off the collections from public view? Why did the university rehouse its anatomical collections in a laboratory complex shared by medicine and the natural sciences? The answer is not obvious, especially not if you started this book by reading the conclusion. In itself, the moving of the anatomy department should not come as a surprise to anyone who has ever read a book about nineteenth-century science and medicine. It fits the standard story perfectly: rise of the laboratory, birth of scientific medicine, expansion of practical teaching. But why would the department pack up thousands of preparations and move them, too? All of that effort to transport fragile jars, skeletons, dried organs – why not leave them behind or throw them out? For the nineteenth-century rise of the laboratory is often depicted not just as a rise, but as a replacement: the lab instead of the museum, experimenting instead of collecting. According to this image, it makes no sense to move the anatomical collections to a laboratory building to use them in the new scientific medicine. As the first two chapters have shown, however, this image is wrong. Plain wrong. Historians have said this before, including Samuel Alberti, Erin McLeary, and Jonathan Reinarz. But besides wrong, the image is also rather persistent; and hence, I say it again: anatomical collections did not disappear in the nineteenth century. They flourished. The rise of the laboratory did not make collections redundant for two reasons. First, the lab did not *replace* the museum – it supplemented it. And second, collections were not tied to the museum – they were used in many spaces and, if you will, in many ways of knowing.

Why, then, does this incorrect image persist? It seems to me that we are misled by the current presentation of historical anatomical collections. The few collections that are easily accessible to the lay public (and that includes historians) all display inertia. We see body parts sealed away in glass jars, neatly arranged on shelves, enclosed in glass cases. The average public anatomy museum screams: 'do not touch!' This supports the idea that anatomical collections are static entities, that the preparations they contain are finished objects, and that audiences are supposed to gaze at the preparations from a safe distance. We find it hard to imagine that things could have been otherwise. Indeed, it is difficult to think of anatomical collections as bustling places when you are standing in, say, the Hunterian Museum's crystal gallery or the specimen hall in the Berlin Museum of Medical History at the Charité, marvelling at the stilled lives all around. But if you study other historical sources in addition to the remaining preparations, as I have done in this book, you will find that, in the nineteenth century, anatomical collections were anything but static. They were moved around, rearranged, extended. Their contents changed continuously, through the acquisition of new preparations, but also through the use and subsequent damage of existing ones. Preparations were meant to be used, and reused, and used again, to be reinterpreted, and to be redissected. Lids were taken off jars; body parts were taken out of the fluid, passed around in class, cut up, and put under the microscope. Anatomical collections and the preparations they contained were dynamic and flexible. Once we understand them as such, it becomes clear that neither the laboratory, nor practical teaching, nor scientific medicine threatened their continued use.

The chapters on students and researchers have shown how both of these audiences used the collections actively. Knowing this, we can understand why the anatomy department in Leiden held on to its thousands of preparations throughout the century. As we saw in the first chapter, anatomical collections suited practical teaching. The collections were needed in teaching laboratories as empirical material: the students required preparations to redissect or to experiment upon. The collections also played an essential role in preparing students for their practical training in the dissection hall: handling preparations helped them learn their facts and handle their feelings. Yet, this continuous need does not fully explain why the anatomy department took *all* of its collections to the new building, including many eighteenth-century preparations. How could these old preparations, based on earlier conceptions of the body and disease, be used in a new medicine? Chapter 2 clarified this, by showing how researchers could continuously reinterpret old preparations because the preparations were made of what they represented. This facilitated the new classification system adopted after the move, for example, for which preparations needed to be reinvestigated and redescribed. In addition to researchers, students also benefited from the flexibility of the preparations: their professors could easily use the old collections to teach them new medicine. Reusing old preparations also saved resources. Replacing old preparations every time medical theories changed would have required more free time and dead bodies than most researchers and teachers had available. Thus, it made perfect sense for the anatomy department to take all its collections to the new laboratory complex.

The ease with which preparations could be reinterpreted might have been a blessing for researchers and students, but it was a curse for non-medical audiences. They lost the collections, and they never really got them back. Today's collections are open to lay visitors in principle, but hard to access in practice. And rather than university governors employing the collections as status symbols, only senior hospital staff members can use the collections to show off the continued excellence of the medical curriculum.

My main story has been that of Leiden's anatomical collections in the nineteenth century, but the dynamic view of anatomical collections and preparations can, and should, be applied to other periods and places as well. Throughout the book, I have drawn on examples from beyond Leiden, most of them from Western Europe, some from the United States. We have seen that other nineteenth-century anatomical collections were handled in similar ways to the Leiden ones. Local variations occurred, of course; recall, for example, the Scottish anatomy teacher Frederick Knox, introduced in Chapter 1, who hesitated to let his students handle preparations because they might gleefully twist toes off skeletons or pierce holes in the covers of preparation jars.[1] The sources I have studied suggest that other British anatomy teachers shared his concerns. They seem to have been more reluctant than their continental counterparts to hand preparations to students. I suspect this reluctance resulted from the British educational system, which relied heavily on private anatomy teachers, especially in the first half of the century. Because they lacked institutional backing, British anatomy instructors had less money and fewer means to build collections than university professors elsewhere. In addition, an insufficient collection implied a direct loss of income: students could and would quickly turn to other, better-equipped teachers. Thus, handling damage had serious consequences – no wonder that many Brits tended to keep students from touching their precious preparations. Further research, in particular comparative histories, would help to test this hypothesis and to identify and explain other differences in the handling of anatomical collections in the nineteenth century.

Such further research should take the idea of anatomical collections as dynamic and flexible entities as its starting point. Local variations notwithstanding, preparations were handled and reinterpreted throughout Europe. Handling practices, especially by researchers, were widespread because they were shaped largely by the universal material properties of the preparations. Preparations could be reused because they were made of what they represented, and this reuse was necessary because the preparations were made of scarce material. (This same scarcity may, as suggested above, have sometimes limited student handling.) Preparations have these material properties regardless of the time and place in which they were made; for although the availability of raw material may vary between countries and centuries, an abundance of body parts of all varieties and pathologies cannot be found anywhere but in utopias (or perhaps, considering the possible causes of that abundance, dystopias). Hence, historians should see anatomical collections as dynamic and flexible regardless of their place or period. They might even benefit from seeing other types of collections as dynamic and flexible as well, especially collections that resemble anatomical ones, such as natural history collections.

This book has explained how anatomical collections ended up where they are now, in the successors to the medical institutions that collected them in the nineteenth century (or even earlier). But along the way, many preparations disappeared. Handling regularly damaged preparations, and parts of the collections were simply used up. Other parts were occasionally thrown out, even if still in reasonable shape. I have stressed how researchers and teachers held on to anatomical collections, but they also, from time to time, discarded individual preparations or even full collections. Shortly after the Second World War, for example, anatomist Julien Fautrez, a professor at Ghent University, destroyed the university's nineteenth-century phrenological skull collection because it had become associated with Nazi ideas on race and eugenics during the war. The trajectory of this collection has been studied by the historians of anatomy Tinne Claes and Veronique Deblon. They show that although the Belgian skulls had repeatedly been adapted to changing research and teaching practices, it proved impossible to shake off their association with Nazi ideology.[2] Researching the disposal of anatomical objects can thus illuminate the limits of reinterpretation; for, as we also saw at the end of Chapter 2, although anatomical preparations are remarkably flexible, the possibilities for reinterpretation are not endless. In addition, studying disposal provides insights into changing attitudes towards collecting and displaying human remains, and Claes and Deblon therefore call for more research on this topic.[3]

This research will be fruitful only if we carefully distinguish between museums and collections; otherwise, we risk concluding that collections disappeared on a much wider scale than they actually did. After the Second World War, the number of medical museums quickly decreased, mainly due to changing teaching practices.[4] A closed museum could spell the end for the accompanying collection, but collections often found other homes. Some museum preparations were incorporated into researchers' handling collections. Others were moved to attics, basements, and closets, where they were kept but became forgotten even by most researchers and students. Eventually, however, all forgotten collections resurface. In 2011, for example, photographer Adam Voorhes visited the University of Texas, where he came across some hundred brains in glass jars, which turned out to have been collected between the 1950s and the 1980s.[5] The brains were still used in medical teaching. In the late 1990s, medical anthropologist Lynn Morgan found dozens of preserved fetuses in a basement at Mount Holyoke College, most of them collected in the first half of the twentieth century.[6]

Such encounters with forgotten collections remind us that, however hidden they may have become, many historical anatomical collections still exist and that, sooner or later, we will have to deal with them. Voorhes photographed the brains he found, and, together with journalist Alex

Hannaford, created a book that introduced the collection to a wider audience.[7] The brains also underwent MRI scanning in preparation for future research use.[8] Most of the fetuses found by Morgan, on the other hand, were destroyed, but not before they had been photographed for the archives.[9] As these two examples show, we can deal with past anatomical collections in different ways. Due to the sensitive nature of the preparations – that is, because they were made out of actual people – it is not always straightforward what way suits which collection. We tend to agree that we should treat human remains 'respectfully', but struggle to define what this 'respect' should entail. Some advocate burying, incinerating, or repatriating human remains in anatomical collections, especially if we know they were acquired without consent.[10] Others feel it may be justifiable to keep such preparations, but that they should not be openly displayed.[11] Yet another group thinks publicly displaying preparations is crucial to facilitate an open, public debate on their history and value.[12]

All these camps have historians in their midst.[13] History cannot tell us the 'right' way to treat human remains. All it can do is offer some clarifications that may help us in our moral reasoning. It shows us how the current situation emerged: studying nineteenth-century anatomical collections has taught us why so many anatomical collections are still kept in medical institutions, hidden from public view. History can also teach us about how anatomical collections function. Of particular importance for the present-day management of historical anatomical collections is the difference between open in principle and accessible in practice. History does not prescribe whether we should publicly display our remaining anatomical preparations, but it does teach us that a principled decision on this matter will count for little unless we pay attention to practicalities such as location, opening hours, and the ease with which doors open and close.

Furthermore, history reveals how the flexibility of preparations combined with the scarcity of anatomical material tempts medical professionals to cling to the anatomical objects they possess and to collect new ones whenever they can. This insight enriches our understanding of the problems posed by historical anatomical collections. It helps explain why medical institutions may hesitate to part with their anatomical collections even if the preparations seemingly serve no direct scientific purpose. This hesitation frustrates advocates for the repatriation of colonial human remains. At the end of her book on nineteenth-century human dissection in England and Tasmania, medical historian Helen MacDonald reproaches the British Museum for holding on to Aboriginal skeletal material 'in the interest of "research" that seems endlessly deferred'.[14] MacDonald depicts this as a convenient excuse that we should not take seriously, but to me it seems that the British Museum is following the medical logic of holding on to flexible and scarce material. Repatriation may nevertheless be the right thing to do, but it will be easier to settle that ethical debate if both sides appreciate each other's arguments.

The promise of future usefulness is also a crucial aspect of the ethical issues raised by contemporary collections of bodily material, such as brain banks, blood banks, and frozen embryos.[15] Most of us agree that donors should give at least implicit consent for the use of parts of their bodies, but non-consensual collecting does nonetheless happen. The illegal harvesting of children's organs at Alder Hey hospital in Liverpool, mentioned in the Introduction, is an infamous example from recent history. Most of the organs harvested by Alder Hey staff were kept in the hospital basements. Some of them were used in research and teaching, but many were not. In particular, the body parts and fetuses collected between 1988 and 1995 under the supervision of the pathologist Dick van Velzen were hardly used.[16] The apparent purposelessness made the situation worse for many parents. As one parent put it during the Parliamentary Inquiry:

> Studying her brain would help explain why her brain did not form properly and it might help treat the next child born with a similar condition. Unfortunately her brain has not been studied. Instead it sits in a jar in a storeroom somewhere.[17]

This parent's anger relates to the anger felt by Helen MacDonald towards the British Museum: why collect or keep body parts if you are not going to use them in any way? The apparent purposeless of such collecting may lead us to dismiss the collectors as immoral, but this will not help prevent such practices in the future. Understanding their motives will – or at least it is the first step.

I think that historians could be of help here. More than one historian has compared the Alder Hey scandal to how nineteenth-century anatomists, especially British ones, appropriated pauper bodies for dissection with little thought for the wishes and fears of the deceased and their relatives.[18] But there is another historical parallel that I think needs to be drawn, one that serves less to condemn and more to understand. What happened in Alder Hey reminds me of the continuous reuse of old preparations, as illustrated by the nineteenth-century afterlife of the Brugmans collection or the early twentieth-century microscopic re-examination of John Hunter's freemartin preparations, both discussed in Chapter 2. These historical examples show that bodily material can be stored for long periods of time (almost a century and a half, in the case of Hunter's preparations) and then become useful in ways never imagined when the material was first collected. This keeps happening in the history of anatomical collections. In 2010 and 2011, for example, researchers from Indiana University extracted DNA from brain tissue collected in the early twentieth century.[19] A psychiatry professor at the same university then used their results in his work on genetic predisposition for mental illness.[20] Even more recently, four Dutch researchers reinvestigated all historical teratological preparations in the Leiden anatomical collections. Early 2018, they reported their findings in an article, in which they also stressed the value of these preparations for current medical research.[21] Thus, holding on to bones, organs, tissue, and fetuses that are not used straight away makes more sense than one would initially think.

Possible reuse is also relevant to material that has been collected with consent. Such consent is often implicit, especially for material taken from people who are still alive. In her first novel, *Man Walks into a Room*, Nicole Krauss describes such collecting. The main character in the book, Samson Greene, is operated for a brain tumour. After the operation, the hospital keeps the removed tissue until, a year later, Samson returns to the hospital – drunk – to demand his tumour back. In the pathology lab, the following dialogue unfolds:

> [The lab technician] backed up against the counter. 'We don't keep it that long,' she whispered.
>
> 'What do you mean you don't keep it? Why don't you keep it?'
>
> 'The tissue disintegrates. We throw it away after a few weeks. We keep a small piece in paraffin. And the slides, those we keep. Those we keep, basically, forever.'
>
> Samson struggled with the idea of his tumor disposed with the rest of the hospital's bloody trash, bone chips and butchery, used syringes and cruddy bandages. ... But there were the slides ... and he would have to content himself with those.[22]

Samson settles for the slides:

> 'Give me my slides,' he repeated.
>
> She had wet, black pupils, the eyes of a small woodsy animal. Her teeth were large. When her mouth was at ease the front teeth strayed rabbit-like below the upper lip.
>
> 'I can't', she said, the lip quivering.
>
> 'But you can,' he assured her ... 'They belong to me.'[23]

Many of us, like Samson, probably have some of our tissue stored in a medical collection somewhere – a blood sample perhaps, a piece of foreskin, a kidney stone, or cervical cells taken out with a Pap smear. In 2009, Dutch medical centres stored 50 million pieces of tissue collected from 14 million individuals, to a population of 16.5 million people (but some of the donors had already died).[24] Most of us agreed, implicitly or explicitly, to the use of the material we donated in all future research and teaching; in other words, we have given broad consent. To determine whether we consider it ethical that many biobanks use broad consent models instead of more specific types of consent, and to determine whether we want to give this consent for our own or our relatives' bodily material, it helps to know the long-term consequences of consenting. Krauss describes how living donors care about how the material is treated *after* it has been collected; this matters to the family members of deceased donors as well, as the reactions of the Alder Hey parents show.

What does the history of nineteenth-century anatomical collections teach us about the future life of the bodily tissue collected nowadays? The answer should be clear by now: that medical institutions will be tempted to hold on to it for many years to come. This is not necessarily a bad thing: the scene from Krauss's novel shows that it may be more disturbing if the material you donate is disposed of – Samson is upset when he learns that part of his body has become medical waste. It is not necessarily a good thing either: it means that our bodily material will be available for use in medical practices and procedures we cannot possibly foresee – the only thing we do know is that the samples and preparations collected nowadays will be flexible enough for new research questions or technologies to be applied to them.

We tend to understand the objects in contemporary tissue collections as artefacts. This suggests that it is technology that renders them immortal – without chemical treatment, Samson's brain tissue will disintegrate. Indeed, biotechnology prevents decay, but immortality requires more: ongoing relevance. The objects in human tissue collections have the relevance they do precisely because they are not fully artefactual – for although they are made, they are made of what they represent, be it cells, embryos, or diseased organs. This specific material property creates an eternal promise of future usefulness. And because of this promise, the blood, brains, and bodies that are currently being collected will be kept, not just for now, or for the coming decades, but, as nineteenth-century anatomical collections teach us, for the coming centuries – or perhaps even longer. As Krauss's lab technician puts it: 'And the slides, those we keep. Those we keep, basically, forever.'

Notes

1 Knox, *Anatomist's Instructor*, 3.
2 Claes and Deblon, 'When Nothing Remains'.
3 Claes and Deblon, 352.
4 On disappearing institutional anatomical museums, see for example Alberti, *Morbid Curiosities*, 197–204; Reinarz, 'Age of Museum Medicine', 432–36.
5 Voorhes and Hannaford, *Malformed*.
6 Morgan, 'Materializing the Fetal Body'.
7 Voorhes and Hannaford, *Malformed*.
8 Voorhes and Hannaford, 168.
9 Morgan, *Icons of Life*, 231; see also Morgan, 'Rise and Demise'.

10 See for example Doyal and Muinzer, '"Irish Giant"'.
11 See for example Engberts and Hogendoorn, 'Een onrustig bezit'; Hon, 'Human Dignity'.
12 Claes and Deblon, 'When Nothing Remains', 359.
13 For opposing views on keeping and displaying historical preparations among medical historians, compare for example Morgan, *Icons of Life*, 224–46, with MacDonald, *Human Remains*, 183–89, and Richardson, *Death, Dissection and the Destitute*, 416–18.
14 MacDonald, *Human Remains*, 184.
15 On the similarities between historical and contemporary collections of bodily material, see also Tybjerg, 'Bottled Babies'.
16 Redfern, Keeling, and Powell, *The Royal Liverpool Children's Inquiry*, 4.
17 Redfern, Keeling, and Powell, 19.
18 See for example Alberti, *Morbid Curiosities*, 205; Richardson, *Death, Dissection and the Destitute*, 415–16.
19 Niland et al., 'High Quality DNA'.
20 Voorhes and Hannaford, *Malformed*, 167–68.
21 Boer et al., 'History and highlights'.
22 Krauss, *Man Walks into a Room*, 186–87.
23 Krauss, 187.
24 Geesink and Steegers, *Nader gebruik*, 13.

Bibliography

Alberti, Samuel J. M. M. *Morbid Curiosities: Medical Museums in Nineteenth-Century Britain*. Oxford: Oxford University Press, 2011.

Boer, Lucas L., Peter L. J. Boek, Andries J. van Dam, and Roelof-Jan Oostra. 'History and highlights of the teratological collection in the Museum Anatomicum of Leiden University, The Netherlands'. *American Journal of Medical Genetics* 176 (2018): 618–37. https://doi.org/10.1002/ajmg.a.38617.

Claes, Tinne, and Veronique Deblon. 'When Nothing Remains: Anatomical Collections, the Ethics of Stewardship and the Meanings of Absence'. *Journal of the History of Collections* 30(2018): 351–62. https://doi.org/10.1093/jhc/fhx019.

Doyal, Len, and Thomas Muinzer. 'Should the Skeleton of "the Irish Giant" Be Buried at Sea?' *British Medical Journal* 343(2011): d7597. https://doi.org/10.1136/bmj.d7597.

Engberts, Dirk, and Pancras Hogendoorn. 'Een onrustig bezit'. *Museumtijdschrift* 23, no. 7(2010): 26–27.

Geesink, Ingrid, and Chantal Steegers. *Nader gebruik nader onderzocht: Zeggenschap over lichaamsmateriaal*. The Hague: Rathenau Instituut, 2009. www.rathenau.nl/nl/p ublicatie/nader-gebruik-nader-onderzocht-zeggenschap-over-lichaamsmateriaal-0.

Hon, Kam Lun. 'Human Dignity and Rights beyond Death'. *Journal of Medical Ethics* 39(2013): 651. https://doi.org/10.1136/medethics-2012-100826.

Knox, Frederick. *The Anatomist's Instructor, and Museum Companion: Being Practical Directions for the Formation and Subsequent Management of Anatomical Museums*. Edinburgh: Black, 1836.

Krauss, Nicole. *Man Walks into a Room*. London: Penguin Books, 2007. First published 2002 by Doubleday.

MacDonald, Helen. *Human Remains: Dissection and Its Histories*. New Haven: Yale University Press, 2006.

Morgan, Lynn M. 'Materializing the Fetal Body, or, What Are Those Corpses Doing in Biology's Basement?' In *Fetal Subjects, Feminist Positions*, edited by Lynn M. Morgan and Meredith Wilson Michaels, 43–60. Philadelphia: University of Pennsylvania Press, 1999. https://doi.org/10.9783/9781512807561-004.

Morgan, Lynn M. 'The Rise and Demise of a Collection of Human Fetuses at Mount Holyoke College'. *Perspectives in Biology and Medicine* 49(2006): 435–51. https://doi.org/10.1353/pbm.2006.0043.

Morgan, Lynn M. *Icons of Life: A Cultural History of Human Embryos*. Berkeley: University of California Press, 2009.

Niland, Erin E., Audrey McGuire, Mary H. Cox, and George E. Sandusky. 'High Quality DNA Obtained with an Automated DNA Extraction Method with 70+ Year Old Formalin-Fixed Celloidin-Embedded (FFCE) Blocks from the Indiana Medical History Museum'. *American Journal of Translational Research* 4(2012): 198–205. www.ncbi.nlm.nih.gov/pmc/articles/PMC3353536/.

Redfern, Michael, Jean W. Keeling, and Elizabeth Powell. *The Royal Liverpool Children's Inquiry: Report*. London: The Stationery Office, 2001.

Reinarz, Jonathan. 'The Age of Museum Medicine: The Rise and Fall of the Medical Museum at Birmingham's School of Medicine'. *Social History of Medicine* 18 (2005): 419–37. https://doi.org/10.1093/shm/hki050.

Richardson, Ruth. *Death, Dissection and the Destitute: The Politics of the Corpse in Pre-Victorian Britain*. 2nd ed. Chicago: University of Chicago Press, 2000.

Tybjerg, Karin. 'From Bottled Babies to Biobanks: Medical Collections in the Twenty-First Century'. In *The Fate of Anatomical Collections*, edited by Rina Knoeff and Robert Zwijnenberg, 263–78. The History of Medicine in Context. Farnham: Ashgate, 2015.

Voorhes, Adam, and Alex Hannaford. *Malformed: Forgotten Brains of the Texas State Mental Hospital*. Brooklyn, New York: powerHouse Books, 2014.

Index